Broadband Data Communications and Local Area Networks

Broadband Data Communications and Local Area Networks

R. G. Willson and N. Squibb

COLLINS
8 Grafton Street, London W1

Collins Professional and Technical Books
William Collins Sons & Co. Ltd
8 Grafton Street, London W1X 3LA

First published in Great Britain by
Collins Professional and Technical Books 1986

Distributed in the United States of America
by Sheridan House, Inc.

British Library Cataloguing in Publication Data
    Willson, R. G.
    Broadband data communications and local
    area networks.
    1. Local area networks (Computer networks)
    I. Title   II. Squibb, N.
    004.4'8   TK5105.7

ISBN 0-00-383095-0

Printed and bound in Great Britain by
Mackays of Chatham, Kent

Contents

Preface

This book presents an introduction to *Broadband Data Communications*. The book began as a set of lecture notes generated for persons who attended the broadband data communications courses given by Nigel Squibb and myself starting in 1982. Although broadband local area networks were introduced into the marketplace in the early 1980s there were, at that time, no established courses available whereby attendees could be introduced to the technology. Nigel and I started these courses to fulfil this market need.

This book introduces the reader initially to the general topic of data communication networks and how these have evolved to satisfy the user needs. The book then goes on to illustrate how broadband local area networks, based on standard cable TV techniques, may be used to satisfy many of the user requirements.

The reader is then introduced to cable TV technology and its application to broadband local area networks. A through understanding of this section would enable the reader to implement a broadband cable system that could host a number of user services.

No technical textbook would be complete without a description of a practical implementation of a broadband local area network. Nigel and I have chosen to described the Sytek LocalNet 20 as a good example of a commercially available broadband local area network. We have also worked closely with each other over the last few years to implement a number of large local area networks in the United Kingdom with LocalNet 20s and therefore feel we have had enough practical experience with this system to be able to present the details for others to read.

The last two chapters of this book describe two particular applications for which broadband data communications networks are ideally suited. The first application illustrates how broadband techniques may be used to distribute and access financial information services. The second illustrates the use of cable TV for the provision of an integrated services network to a local community. This example was described in a paper given by Bernard Clements of Ewbank Preece Management Ltd at the Cable 85 Conference and is reproduced here with his permission.

The material in this book has been gathered from many sources.

We must however thank Sytek Inc. for allowing us to include information available to us in the form of technical manuals, presentation material, seminar and user group meeting notes. We would in particular like to thank the following members of Sytek for their assistance, advice, comments and discussion on many points – Greg Ennice, Peter Filice, John Corey, Ed Cooper, Ken Biba, Phil Edholm and Jerry Yost.

We hope the material presented here will assist managers, designers, implementors and operators involved with data communications to understand the principles and techniques of broadband networks. The reader should have gained an understanding of the capabilities and limitations of the broadband approach and appreciate the task undertaken by the broadband system designer.

Welcome to *Broadband Data Communications*.

R. G. Willson

CHAPTER 1

Introduction to Networks

The modern world uses many electrical and electronic devices to convey information between human beings. These devices are located at distances such that direct human contact is not feasible. The primary forms of information used by humans to communicate are:

Voice.
Text.
Image.

The electrical or electronic devices used to convey the above types of information are:

Telephone.
Computer.
Television.

With the increasing use of the above devices it becomes necessary for larger scale interconnection of these devices to meet user demands for an increasingly more sophisticated service.

The telephone system grew from the local interconnection of telephones in a town to a worldwide, automatic, subscriber dialled network which may use communication satellites to reach the far corners of the earth.

Computers were initially installed to provide a specialised calculation facility at a local site. The modern computer today ranges from a personal computer at a user's desk to a 'super' computer housed in an air conditioned building that may be used for forecasting the weather. Computers are also increasingly interconnected to form networks. Networks allow users to use a simple terminal or a personal computer connected to more powerful computing resources located elsewhere. Powerful computing resources can therefore readily be made available to a user at his or her desk.

A network of computers is not used only to make powerful computing resources available to a user at his or her desk, but allows the interconnection of much less powerful computing devices. Devices such as personal computers may be used for the transmission and reception of information at the personal level, e.g. mail. Information may be created and stored by an individual at a personal computer, and this information made available to co-workers using a network.

Television or the broadcasting of moving images with accompanying sound is now potentially available as sets of national broadcasting networks that can be interconnected via the medium of direct broadcast via satellite (DBS). Instant television news is flashed across the world via satellites so that we can now view events live as they happen anywhere on earth or in near space.

Cable television networks were originally developed to get good television images to communities in areas where the broadcast signals were weak. Cable television has now developed into a cable broadcasting service that allows the subscriber additional choice of educational or entertainment programmes, many of which are generated locally and appeal to specialist audiences.

Each form of information has developed a specialised network for the transmission and reception of information together with the appropriate control and management required to provide the necessary service to its subscribers. With the increasing use of digital techniques, not only for the transmission and reception of all forms of information but also for the manipulation, transformation, storing and processing of this information, there is a natural convergence of techniques taking place. Simultaneously as digital techniques permeate all forms of information processing, the volume of information, and hence the speed or bandwidth required to convey that information, are increasing dramatically. Networks are therefore becoming necessary when large amounts of information, coded in digital form, are required to be transported, switched and processed at high speed.

Cable television networks are well developed, wide bandwidth distribution networks. This book will discuss how cable television distribution networks can be used to provide a broadband data communication facility over a local area.

Before we examine the use of cable television distribution networks for broadband data communications, we shall briefly look at the types of networks currently implemented so that broadband networks covering a local area can be seen in their proper perspective.

Networks may be classified into three broad categories:

Wide area networks (WAN).
Metropolitan area networks (MAN).
Local area networks (LAN).

WIDE AREA NETWORKS (WAN)

Wide area networks are networks with national and international coverage. These networks are normally implemented by national organisations such as the post, telegraph and telephone (PTT) authorities. Via international agreement between national bodies, interchange of traffic between countries is achieved.

Existing wide area networks provide facilities in the UK for the transmission and reception of the services shown in Table 1.1. Because wide area networks do interconnect over large distances it is expensive to allocate large amounts of bandwidth for each connection or circuit. A typical circuit parameter in a wide area network would be:

Distance $>10^3$ km
Bandwidth 4 kHz–10 MHz
Data rate 10^4–10^6 bps
Delays 300 ms

Table 1.1

Information service	Networks
Image	Broadcast TV/BBC, ITV, video conferencing.
Data/text	Public data networks, e.g. telex, packet switched stream (PSS). Private data networks, e.g. IBM SNA networks, ICL IPA networks. Leased lines/datel service.
Voice/data	Public switched telephone network (PSTN). Private telephone network

METROPOLITAN AREA NETWORKS (MAN)

Metropolitan area networks are networks which provide city wide coverage. These networks are owned and operated by publicly regulated bodies. In the UK, as a result of the privatisation of British Telecom, the establishment of Mercury Communications Ltd to compete with British Telecom and the creation of the Cable Authority to award cable francises, the development of metropolitan area networks to provide new and innovative services is poised for exploitation by private investment. Whether the potential rewards will encourage private investors to take the investment risks only time will tell. This book describes technology that is currently available and which could enable many of the proposed new services on wideband cable networks to be implemented.

Metropolitan area networks provide the information services shown in Table 1.2. Typical network parameters for metropolitan area networks are:

Distance <20 km
Bandwidth 10 kHz–50 MHz
Data rates 10^3–10^7 bps
Delays 10 ms

Table 1.2

Information service	Networks
Image	Cable TV, video conference.
Data/text	Cable TV (broadband data communications)
	Public data network
	Telex
	Packet switched stream (PSS)
	Private data network
	Leased lines/datel service.
Voice/data	Local telephone exchange
	Private telephone exchange.

LOCAL AREA NETWORKS (LAN)

Local area networks are networks that provide coverage on private sites such as an office building, a manufacturing site or a university campus. In general, no permission from any public regulatory body is necessary to establish and run such a network.

Local area network requirements have resulted from the computer industry's need to interconnect large numbers of computers, workstations and terminals to provide interworking at a single site. In general the computer industry has solved this requirement by providing a specialised, high speed, single shared (multiplexed) channel via which all computers, workstations and terminals are interconnected. This solution has resulted in a single data/text network which is shared by all the computers, workstations and terminals on the local site. This type of local area network (LAN) is commonly known as a *baseband LAN*.

Cable TV technology can also be used to implement networks which provide both an image and data/text information services within the confines of a private site. Cable TV technology allows the implementor to operate multiple information services on the same distribution network. Local area networks using cable TV as the transmission and distribution media are known as *broadband LANs*.

The information services that are shown in Table 1.3 can be provided by a local area network. Typical network parameters for a local area network are:

Distance <10 km
Bandwidths 1–200 MHz
Data rates 10^6–10^8 bps
Delays 1 ms

Figure 1.1 summarises the various types of network.

Table 1.3

Information service	Networks
Image	Cable TV, video conferencing.
Data/text	Cable TV (broadband LAN) Private branch exchange (PBX) Local area network Data PBX Campus packet switch Dedicated lines.
Voice/data	Private branch exchange (PBX).

Local area networks and metropolitan area networks can both be implemented using cable TV transmission and distribution technology. It should however be noted that telephony is used in wide area, metropolitan and local area networks to provide voice and data information services. The maximum data rates currently achievable using this technology preclude it from being extensively used for high speed data communications in metropolitan and local area networks. There are, however, many interesting innovations currently being undertaken where the functions of a telephone exchange and a broadband local area network are being combined to provide an image/data/text/voice service from a single network. This network is known as an *integrated services local network* (ISLN).

- **Wide area network (WAN)**
  Broadcast TV
  Public switched telephone network
  Public and private data networks
  Leased lines $\begin{cases} \text{Voice} \\ \text{Data.} \end{cases}$

- **Metropolitan area network (MAN)**
  Cable TV
  Local telephone exchange
  (Local data network)
  Leased lines.

- **Local area network (LAN)**
  (Image service)
  Private branch exchange
  Local data network
  Dedicated lines.

Fig. 1.1 Summary of the network types.

CHAPTER 2

Integrated Services Local Area Network (ISLN)

The object of an integrated services local network (ISLN) is to provide a user via a single network with all forms of information services – image, data/text and voice.

In order to study the feasibility of integrating services via a single local network we need to appreciate, in broad terms, the techniques currently used to provide services via today's specialised networks. There are four fundamental techniques used to provide information services:

- Broadcasting.
- Circuit switching.
- Private dedicated circuits.
- Packet switching.

BROADCASTING

A broadcasting technique is used where it is necessary to transmit information from a single source to many destinations. An application of this technique is the multi-channel distribution of television programmes either off-air or via a cable TV system.

The same broadcast technique can also be applied to the simultaneous distribution of data from a single source to multiple destinations. If it is required that specific data elements must be sent to identified destinations, this can be achieved by tagging the data elements with their destination addresses and then broadcasting. The destination will receive all information being broadcast, but with additional logic at the destination the destination will discard all data that does not contain the correct destination address, and only select data with the correct destination address.

An application of broadcasting data and allowing selection to be performed at the destination is the simultaneous broadcast of a television programme (image) with teletext pages (text). At the destination, a user can instruct a television receiver only to select a page of data identified with a specific address (see Fig. 2.1).

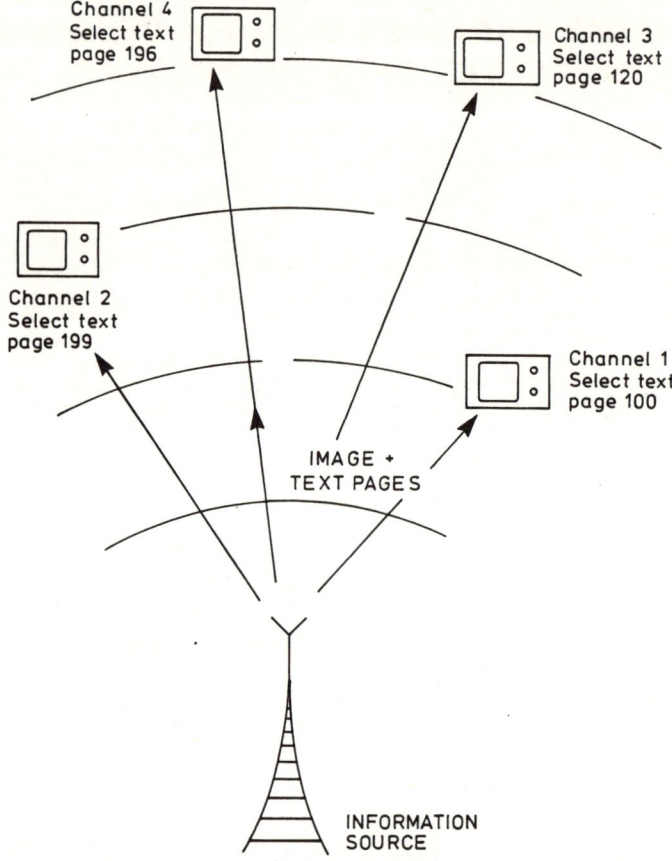

Fig. 2.1 Broadcasting.

CIRCUIT SWITCHING

Circuit switching techniques provide a means of connecting two subscribers via a network where, once connected, the two subscribers have dedicated use of all the bandwidth provided by the network for that particular connection or circuit.

The most common example of circuit switching is the telephone system (see Fig. 2.2). In the telephone system once two subscribers have been connected via the telephone exchange a dedicated path is established between the subscribers. This path or circuit is normally used for voice communication but the same path can be used for transmitting data or text between the two subscribers. In order to achieve data communication via the present telephone network it is necessary to convert digitally encoded data or text into audio tones so that the information can be transported via a network that was primarily designed for voice communication.

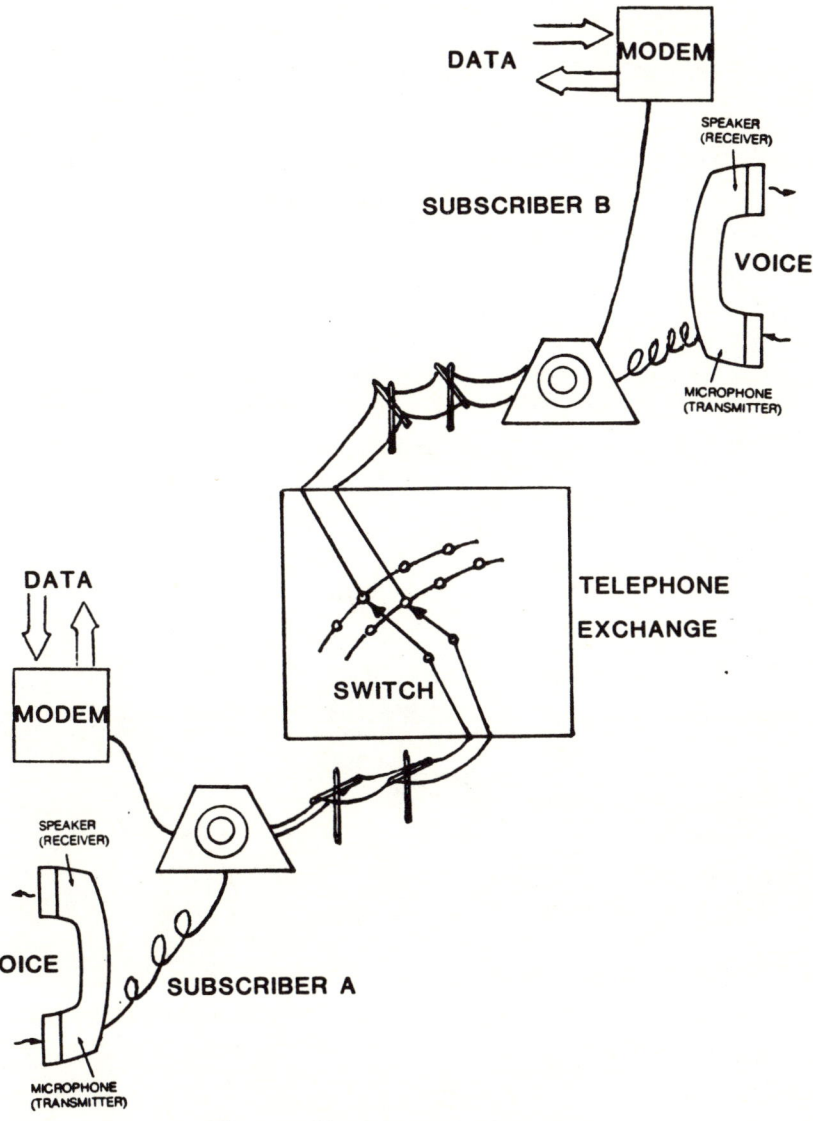

Fig. 2.2 Circuit switched service.

PRIVATE DEDICATED CIRCUITS

Private circuts are special circuits that are always dedicated for the use of two subscribers. These circuits are never switched by the network, but can be used for voice or data traffic. (See Fig. 2.3.)

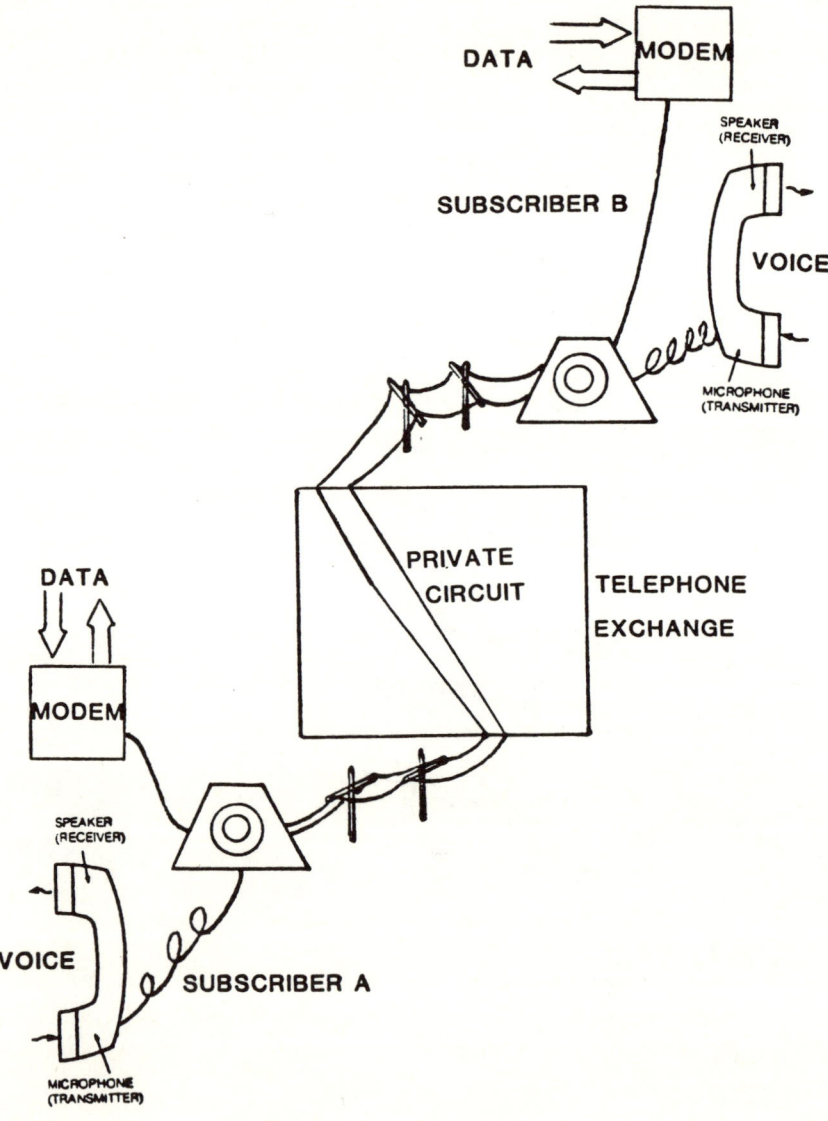

Fig. 2.3 Private circuit.

PACKET SWITCHING

With the increasing use of computers by a large population of terminals, the cost of individually connecting terminals to their appropriate service computers becomes prohibitive. Due to the fact that the data traffic between terminal and computer is bursty and

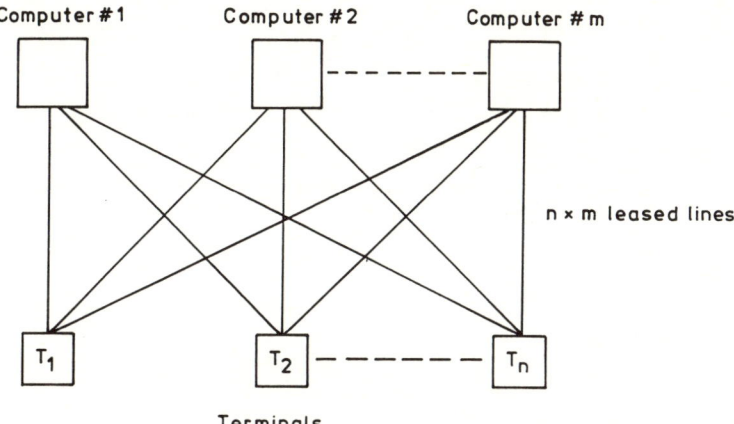

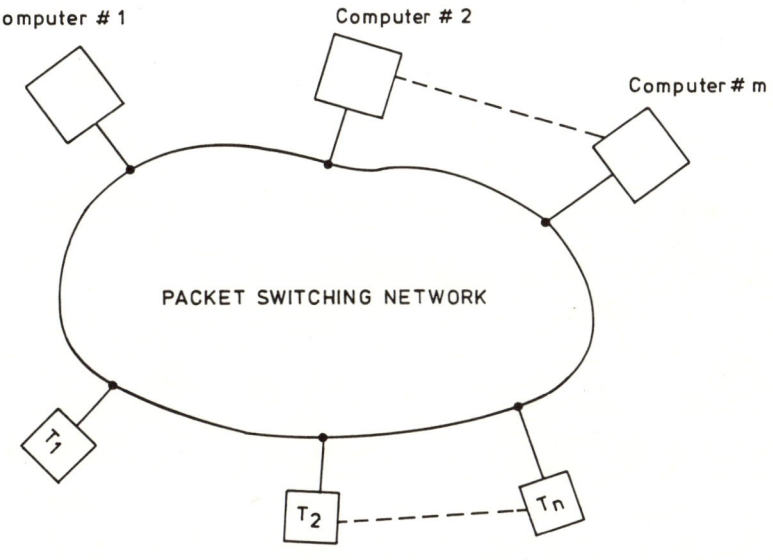

Fig. 2.4 Evolution of packet switching.

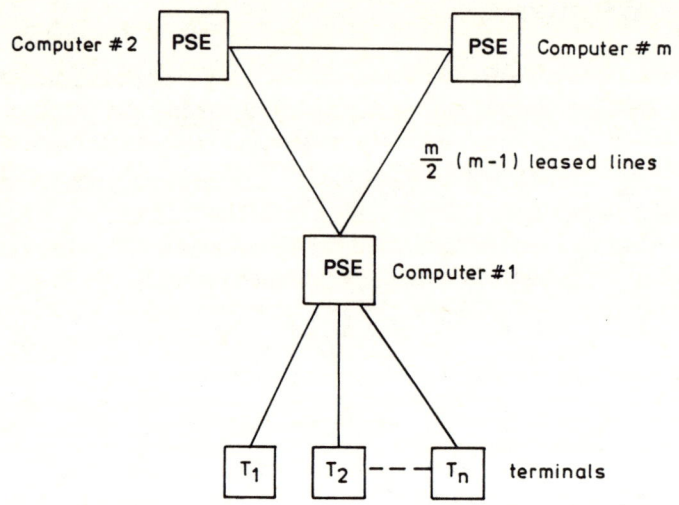

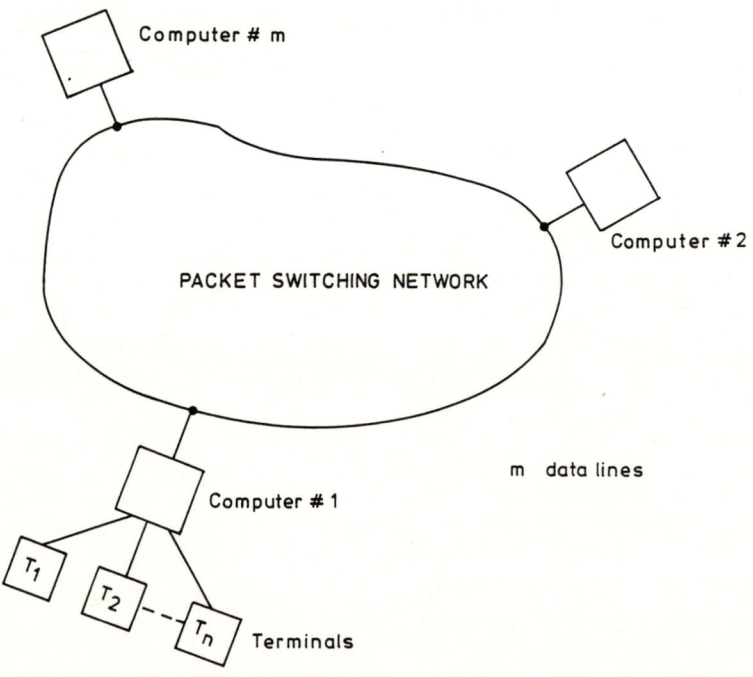

Fig. 2.5 Support for terminals in packet networks.

sporadic by nature, the individual lines connecting terminals and computers are totally under-utilised. More flexible interconnection and better utilisation of the communication lines can be achieved by interposing a packet switching data communications network between the computers and the terminals (see Fig. 2.4).

A packet switching network is provided by the interconnection of *packet switching exchanges* (PSEs) (see Fig. 2.5). This means that instead of adopting the techniques used for telephone calls, where a circuit or transmission path is constantly maintained between the two parties for the duration of the call – circuit switching – in a packet switching network the 'message' is broken into discrete quantities, wrapped separately in control information, and sent through the network as distinct entities called *packets*. The packets are delivered in the correct order at the destination, the control information is discarded and the message is reassembled into its original form.

The advantage of this method is that packets from various sources may be interweaved and therefore share transmission paths as they travel through the network. The transmission paths are therefore more heavily utilised and the traffic load carried with the maximum transmission efficiency.

A set of instructions is required to break a 'message' into blocks and wrap it with the necessary address and control information so that the network can route the packet to its destination. Such instructions are known as *protocols*.

Simple user terminals frequently do not have the computing capability to implement the protocols necessary to interface directly with a packet switching network. Under these circumstances terminals are connected to a local computer in which the networking protocols have been implemented.

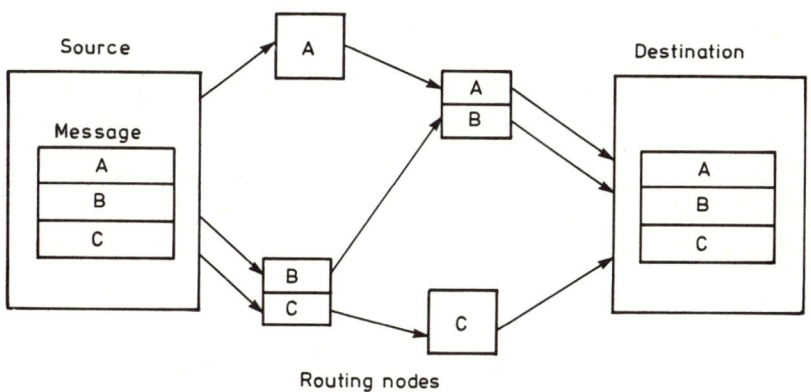

Fig. 2.6 Store and forward.

Where the physical network extends over large distances – wide area networks – packet switching is implemented via the interconnection of packet switching exchanges or routing nodes. These nodes examine the address and control information that is wrapped around each data block that forms a packet and routes the packet, if necessary, via additional nodes to its destination. The packets are

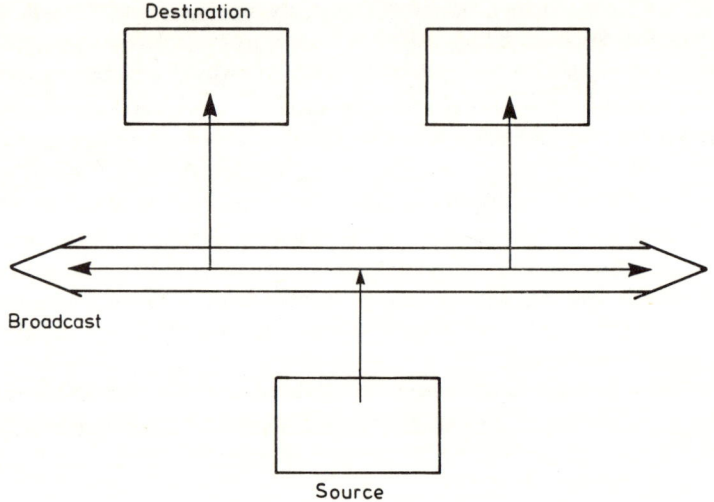

Fig. 2.7 Packet switching.

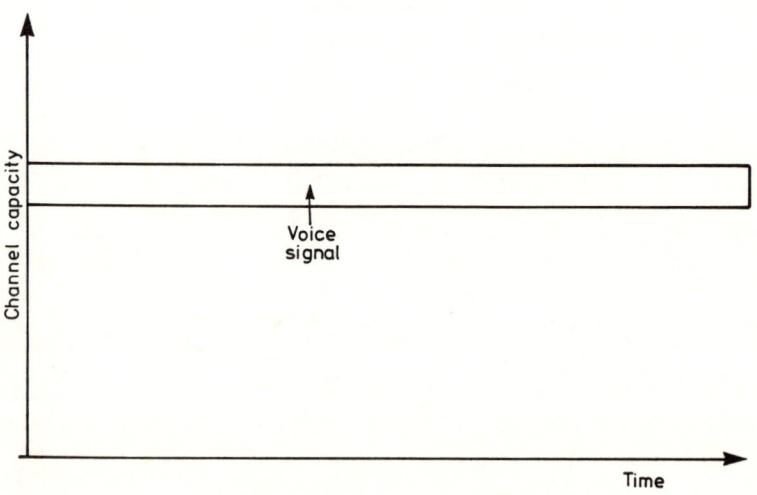

Fig. 2.8 Circuit switching

stored and forwarded via intermediate routing nodes to their final destination (see Fig. 2.6).

A packet switching network can also be implemented over a restricted geographical area - metropolitian area or local area - where the physical transmission path is a single cable system. Because of the restricted geographical area the source node can broadcast a packet into the cable system and this packet will be received by all other nodes attached to the cable system (see Fig. 2.7). The packet will contain data plus the address of the destination node. All nodes receive the packet but only the correct destination node will process the packet any further to recover the data. Packet switching protocols used for wide area networks are therefore very different from those used on metropolitan and local area networks.

In summary, wide area networks implement packet switching by the interconnection of intermediate routing nodes. Packets travel from source to destination via multiple hops; a routing node determines the route to be taken for the next hop. In metropolitan and local area networks, packets are broadcast by the source and are received by all destinations, and packets therefore travel from source to destination in a single hop.

Circuit switching dedicates a proportion of channel capacity to the subscriber connection, this channel capacity being dedicated to that connection until the connection is broken (see Fig. 2.8).

Packet switching uses all the channel or transmission capacity only when a packet is travelling through the system. This occurs

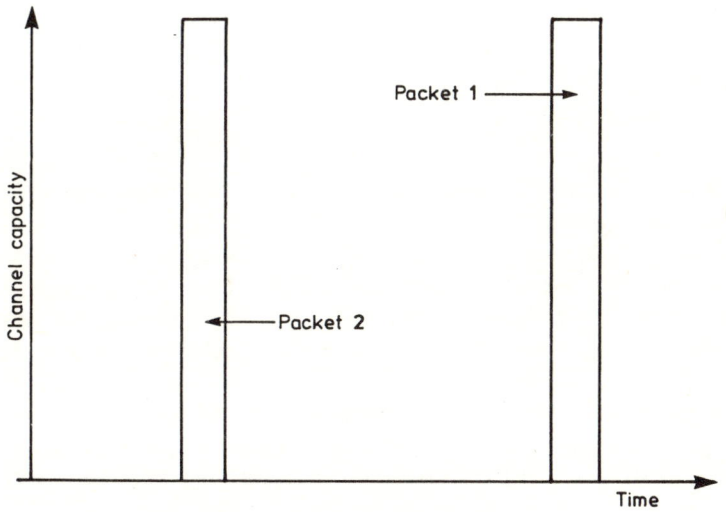

Fig. 2.9 Packet switching.

when packets are being transmitted between routing nodes in a wide area network or when packets are being broadcast on a cable system (see Fig. 2.9).

USER REQUIREMENTS

We have looked at the fundamental techniques to transport and switch information from a source to a destination point. We now need to review the user's requirements for such a network and then determine the basic characteristics for such a network that will satisfy these needs.

A user organisation today has an increasing number of subscriber terminals that transmit and receive information; some of these are shown in Table 2.1. It is therefore true to say that an integrated services local network (ISLN) must be able to accept and deliver multiple information streams to and from subscriber locations within this local network.

Table 2.1

Subscriber terminals	Information services
Telephone	Voice.
Visual display unit	
Data processing computers	
Personal computer or workstation	Data/text.
Facsimile	
Television	Image.
Video camera	

As we have discussed earlier, the various information services – voice, data/text and image – evolved over the years as separate independent networks with architectures to suit (see Fig. 2.10). Users have therefore already incurred major capital expenditure and many subscribers' terminals have been installed and are constantly in use. A user, therefore, would require an ISLN to be evolutionary, so that existing subscriber terminals could be attached to and supported by the new ISLN.

The geographic coverage required for a local area network may therefore vary from providing connections for users within a single department to providing service connections to subscribers across a city.

The network, as we have seen, may be required to accept and

Information	Service
DATA:	Packet switched service Circuit switched service Private circuit.
VOICE:	Circuit switched service Private circuit.
TELEVISION: (image)	Multi-channel distribuiton Circuit switched service.

Fig. 2.10 Types of information and services.

- Terminal-to-host communications
- Workstation-to-workstation communications
- Shared access to special purpose devices (database machines, graphics systems)
- Manufacturing process control
- Environmental control/security
- Communicating word processors
- Electronic mail/voice mail
- Teleconferencing
- Video conferencing
- Office of the future
- Television

Fig. 2.11 Local networking applications.

transport data generated from a variety of subscriber terminal devices. Also, because the network must be evolutionary, existing network architectures must be supported. In practice this may mean that existing point-to-point or multipoint connections need to be maintained across the ISLN. In addition, other subscribers' terminals used for different applications need to be connected dynamically to a variety of destinations by subscriber initiated commands.

Alternatively, users can view an ISLN simply as a convenient mechanism to support a multitude of user applications, some of which are listed in Fig. 2.11.

Local networking user requirements can then be summarised as:

- The network must protect the user's existing capital investment in subscribers' terminals and devices.
- The network must provide a high degree of inter-connectivity.
- The network must provide the user with a high degree of geographic independence. For example, if a user moves to another location, then

by connecting to the network at the new location all information
services previously available to the user should be accessible again.
- New services should be easily added to the network.
- The network should be robust and resilient. Any physical damage
to the network should only affect subscribers that are in the vicinity
of the damaged section. Where necessary the network may be
designed to be resilient and service maintained by by-passing a
failed or damaged section.
- The cost to implement an ISLN must show cost benefits where
compared to implementing separate specialised networks for each
information service.
- Finally, because of the increasingly rapid developments in all areas
of computers, telecommunications and microelectronics, an ISLN
must evolve to take account of new developments. The technology
selected for an ISLN must be 'future proof'; it should not be

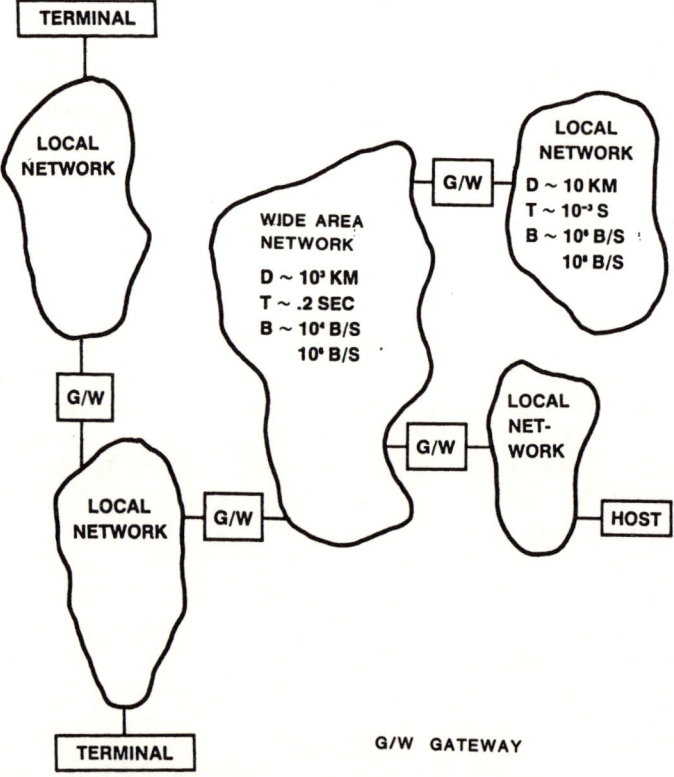

Fig. 2.12 The generalised network.

implemented using technology which is likely to be superseded in the near future.

To satisfy these requirements the generalised network shown in Fig. 2.12. as a solution must contain the following:

- The network must transport simultaneously multiple information streams to support voice, data/text and image.
- The transmission media must be easily accessible to subscriber terminals and devices.
- The transmission media must support wide bandwidths to be able to transport many streams of information of varying bandwidths.
- The local network must be able to support simultaneously multiple network architectures. Network architectures are established via commercial pressures, user acceptance, cost and suitability for the application. No single architecture necessarily provides the user with the ultimate solution.
- Distributed intelligence using a microprocessor will be used to implement the many functions required in an ISLN.
- A user connected to a local network cannot function if the local network exists only as an insular entity. The local networks must be interconnected firstly to each other, and secondly to networks that span larger geographical distances – wide area networks.

CHAPTER 3

Local Distribution Networks

There are in existence currently two major local distribution networks (see Fig. 3.1):

Local telephone system
Cable television.

The local telephone system will transport and switch voice and data traffic. The telephone system was primarily designed to carry voice traffic generated by subscribers' telephones. With the development of *modems* (modulator/demodulator), digital information is transformed into audio tones that are suitable for transmission over the telephone network. This is a circuit switched network. Due to user demand for a specialised network for transporting and switching data only, packet switching exchanges were developed and interconnected by telephone lines using high speed modems to form the backbone of a packet switching network.

As digital techniques become increasingly used by the telephone network for trunk transmission and switching, voice information will be digitised and transported across the network as digital data. Voice as digital data will therefore be indistinguishable from digital data generated by a computer or terminal.

The speed of data transfer is limited for two reasons. First, because the telephone network is primarily a network for voice traffic. This determines the primary data rate for a single digital circuit as 64 kbps. Although a data rate of 64 kbps would be adequate for many terminal-to-host computer applications, it is not sufficiently high for many workstation-to-host computer, workstation-to-workstation and host computer-to-host computer applications.

The second limitation inherent in the existing telephone network today is the copper pairs used to connect subscribers to the local telephone exchange (see Fig. 3.2). The telephone network has a vast investment in copper pairs in the local distribution network, and although in time the long distance trunk circuits may be implemented in fibre optics, it is inconceivable that fibre optics will replace the copper pairs used to connect subscribers to the local telephone

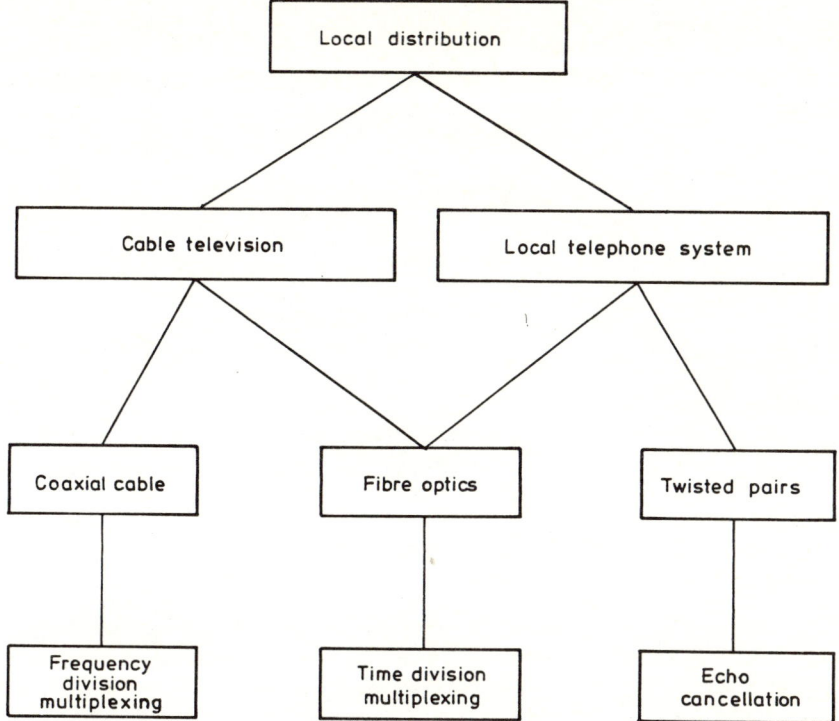

Fig. 3.1 Cable stream.

Telephone cable

Short distance

Low data rate

Not usually shielded

Fig. 3.2 Twisted-pair cable.

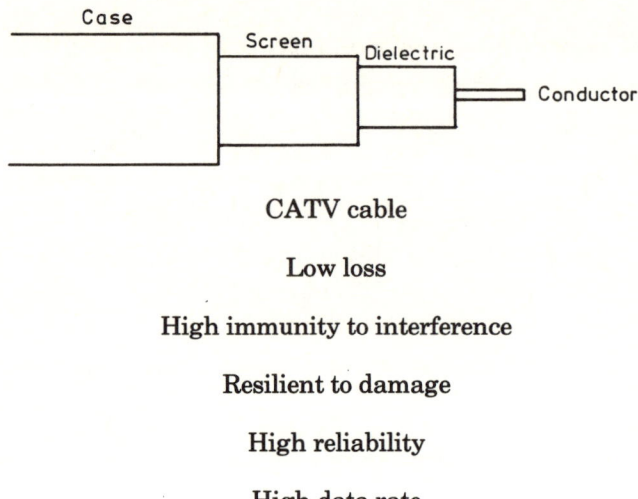

CATV cable

Low loss

High immunity to interference

Resilient to damage

High reliability

High data rate

Fig. 3.3 Coaxial cable.

exchange in the short term. Digital techniques are however being applied to the local copper pair distribution network and this will allow the subscriber to access simultaneously:

2×64 kbps channels, and
1×16 kbps channel.

The maximum data rate available to a subscriber is therefore limited by the characteristics of the local copper pair distribution network. Developments in the telephone network will not easily enable personal computers to be networked with each other at a basic data rate of 2 Mbps. Neither will the local telephone network support host computers to be networked at a data rate of 10 Mbps.

Cable television provides a second local distribution network. It is implemented via coaxial cable (see Fig. 3.3), and was primarily designed to broadcast television to its subscribers. Television is broadcast via the coaxial cable distribution system using radio frequency carriers operating at very high frequencies (50 – 450 MHz, VHF).

Digital data can also be broadcast via the coaxial cable distribution system by the use of radio frequency modems. Radio frequency modems transform digital information into modulated radio signals that are suitable for transmission over the coaxial cable distribution network. The same coaxial cable system can therefore transport simultaneously television (moving images with accompanying sound) and digital data.

To provide a subscriber with a cable television service it is necessary to connect the subscriber to the main coaxial cable distribution system. Each subscriber requiring a cable television service will be connected via a coaxial cable, usually known as the subscriber's *drop cable*, to the main distribution backbone. The subscriber's drop cable is a coaxial cable normally designed to deliver to the subscriber many wideband radio frequency channels of television. The subscriber's drop cable may operate up to 200 MHz or 400 MHz depending on the design of the particular system.

Using radio frequency modems the subscriber's drop cable can in addition to television carry wideband radio frequency channels of data. The subscriber's drop cable provides him with wide bandwidth access to information services. Because of the wide bandwidth available, several information services, each of different bandwidths, may be simultaneously multiplexed to and from the subscriber via the single coaxial cable.

The cable television industry has demonstrated that large city wide networks can be constructed that will distribute up to 30 channels of television. Cable television techniques therefore provide us with a well proven transmission media that can be applied to the transmission and distribution of data.

Coaxial cable provides the designer with an encapsulated electromagnetic transmission media that will distribute wide bandwidth radio frequency energy. Information modulated on the radio fre-

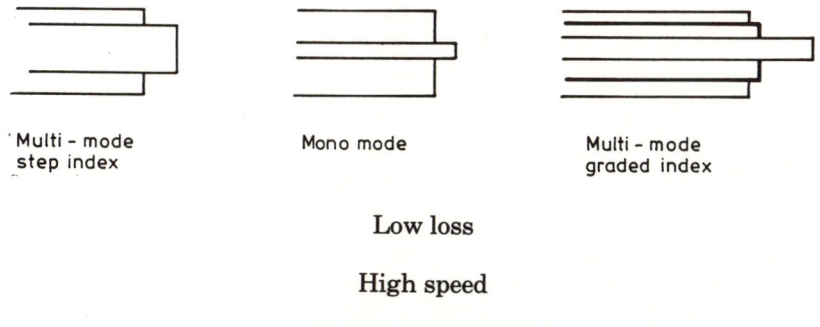

Multi – mode step index   Mono mode   Multi – mode graded index

Low loss

High speed

Immune to interference

Two-way transmission requires a pair of fibres

Point-to-point circuits

Unsuitable for broadcast bus sytems

Fig. 3.4 Fibre optic cables.

quency carriers can therefore be transmitted and received via this media. By using standard cable components and amplifiers a cable distribution system with a bandwidth of 450 MHz can readily be implemented. The cable system can therefore support simultaneously many 10 Mbps data channels.

Fibre optic cable provides a transmission media with some attractive properties (see Fig. 3.4). It provides a low loss path to high speed signals and is immune to electromagnetic interference. Fibre optic cable, however, does not currently provide a cost effective method to access or 'tap' the cable at frequent intervals, as required in a distribution network, to send and receive information. Current techniques require optical-to-electrical interface at each tap point and this is not economic in a large network that may consist of many thousand tap points. Fibre optics does currently provide a good transmission media for long point-to-point circuits but is not suitable for a broadcast bus system that requires frequent access or tap off points.

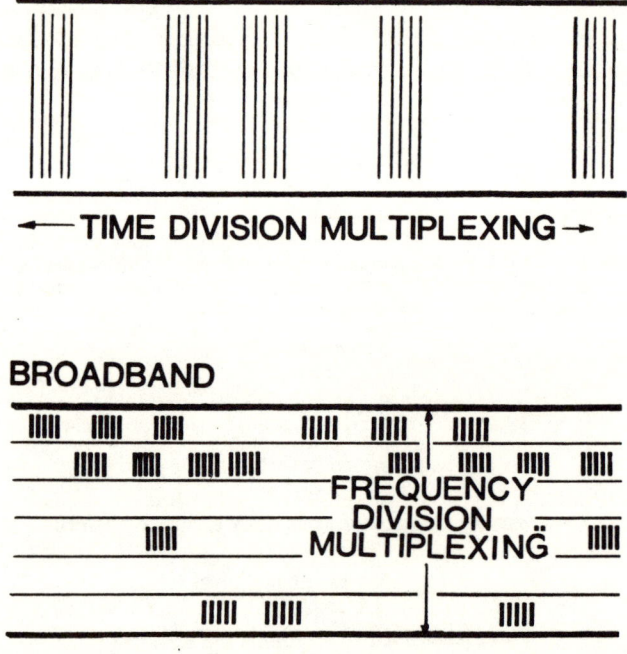

Fig. 3.5 Broadband gives two-dimensional multiplexing.

Information exists as a time varying electrical signal. The amount of information is proportional to the frequency bandwidth of this signal. If this time varying signal is carried directly by a cable system it is termed a *baseband transmission system*. A baseband system therefore transmits a single high speed information channel on a cable system. The single high speed information channel can be used or shared by many devices by interleaving information from the various devices on to the cable at different times. The technique is known as *time division multiplexing* if the time intervals are

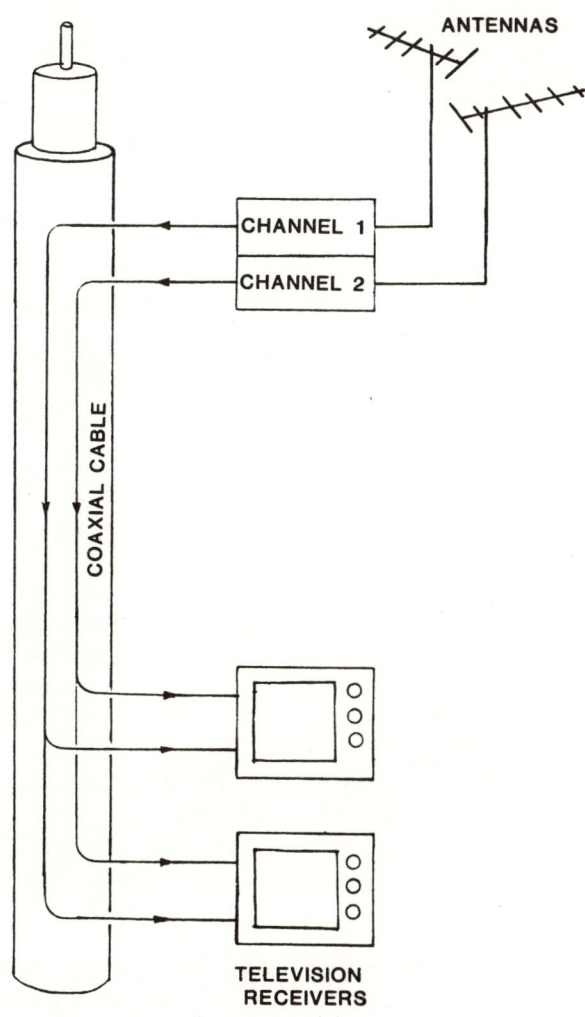

Fig. 3.6 Multi-channel TV distribution.

regularly spaced (see Fig. 3.5). If the information channel is shared by many devices at random intervals this technique is known as *statistical multiplexing*.

If the time varying signal representing information modulates a radio frequency carrier, and this radio frequency carrier is transmitted via the cable system, this is known as a *broadband system* (see Fig. 3.5). By modulating sources of information at different carrier frequencies, all the information streams can be simultaneously transmitted via a single cable system. This technique is known as *frequency division multiplexing*.

Each information stream can be time divisioned independently or statistically multiplexed to support many devices. Broadband therefore provides the designer with two dimensions of multiplexing. The broadband system enables a given cable to carry and distribute a large volume of information.

CABLE TV FOR DATA NETWORKS

Coaxial cable distribution systems designed for cable TV can provide an excellent transmission media for the construction of wide bandwidth data networks.

Frequency division multiplexing

A coaxial cable distribution system implemented via the use of standard cable TV components will provide a network designer with 450 MHz of bandwidth with a geographical coverage represented by 25 miles of cable.

Television is a mechanism by which we code a radio frequency signal with information that represents visual images and sounds. This radio frequency signal is normally broadcast via a high power transmitter. A television aerial together with an appropriate receiver will allow a user to receive the signal, decode the information and reproduce the visual information and sounds.

Each television service will broadcast at a radio frequency allocated for the particular service. Therefore, a number of television services or programmes can be broadcasting simultaneously and because they operate at different frequencies there will be no interference at the television receiver between the different television programmes.

The subdivision and allocation of the radio frequency spectrum to the various services that require to operate simultaneously is known as *frequency division multiplexing*.

Coaxial cable is also an electromagnetic media with a defined radio frequency spectrum. Frequency division multiplexing is used to allocate the necessary bandwidth to the various information services that require simultaneous use of the cable.

TV distribution

Figure 3.6 illustrates how multiple television programmes that are received from off-air broadcast are distributed via a coaxial cable

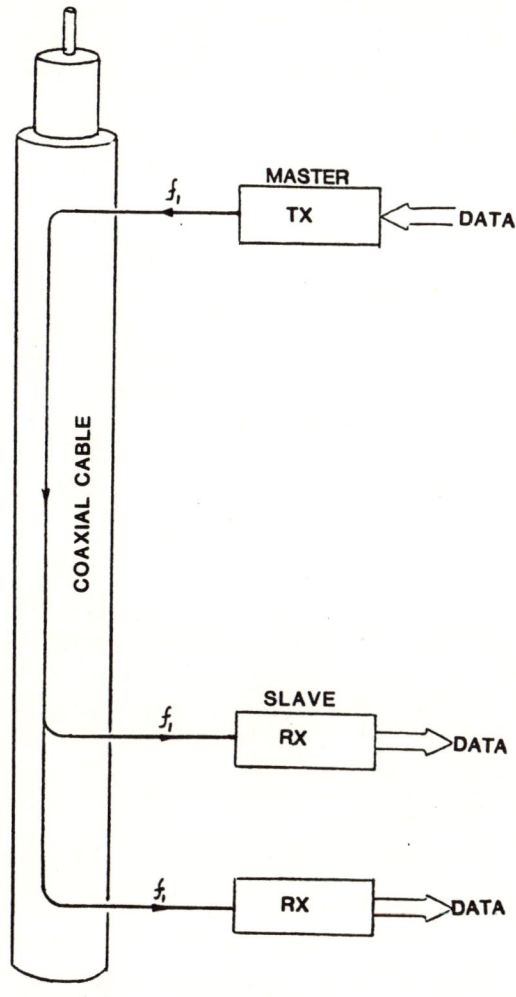

Fig. 3.7 One-way broadcast of data.

system to individual subscribers' television receivers. Each off-air television programme is broadcast at a particular radio frequency on the cable system. Subscribers connected to the cable system may view the appropriate programme by tuning their television receivers to the correct frequency or channel.

One-way broadcast of data

Television may be considered to be a one-way broadcast of information. If the original information was not coded signals representing something visual, but binary information representing text or data, this information could also be broadcast over the cable system (see Fig. 3.7).

The binary information is coded and modulates a radio frequency transmitter. This signal is broadcast at a defined carrier frequency over the cable system.

Receivers located anywhere on the cable system and tuned to the correct frequency will receive the binary information in coded form. Decoding or demodulating the signal will recover the original text or data, which can then be reproduced on an appropriate device.

The two processes, *modulation* and *demodulation*, constitute a device known as a *modem*. Modems are frequently used to transmit and receive data via the telephone network. Voice frequency modems code binary information into signals that are suitable for transmission and reception over the telephone network. Radio frequency (RF) modems code binary information into signals suitable for transmission and reception via a coaxial cable system.

Coaxial cable systems are systems of wide bandwidth and RF modems can be implemented to accommodate data transmission at high speed, at a variety of data rates, as shown in Table 3.1.

We have up to now considered only the one-way broadcast of data. Interactive systems and in particular computer systems require both the transmission and reception of information between terminal and a computer system or between workstations or computers themselves. We therefore require to implement a two-way data communication system.

Two-way data communications

Simultaneous two-way communication between two subscribers A and B can be achieved if each subscriber is equipped with both a transmitter and a receiver. Subscriber A's transmitter is set to transmit at frequency f_1 and his receiver is tuned to receive at

Table 3.1.

	Data rate	*Bandwidth*
Low speed	up to 19.2 kbps	30 khz
Medium speed	64 kbps	300 khz
High speed	2.048 kbps	6 khz

frequency f_2, while subscriber B is set to transmit at frequency f_2 and receive at frequency f_1. By the allocation of two frequency channels, simultaneous two-way communication can be achieved between two subscribers.

This concept of two-way communication needs to be generalised so that in an interactive computing environment many terminals can access a computing or data processing service and interact with the appropriate application programs.

A cable system will carry signals travelling in either direction. A typical cable system topology, commonly known as *tree and branch*, is shown in Fig. 3.8. The start of the cable system is known as the *headend*. In a cable TV system the TV programmes are broadcast

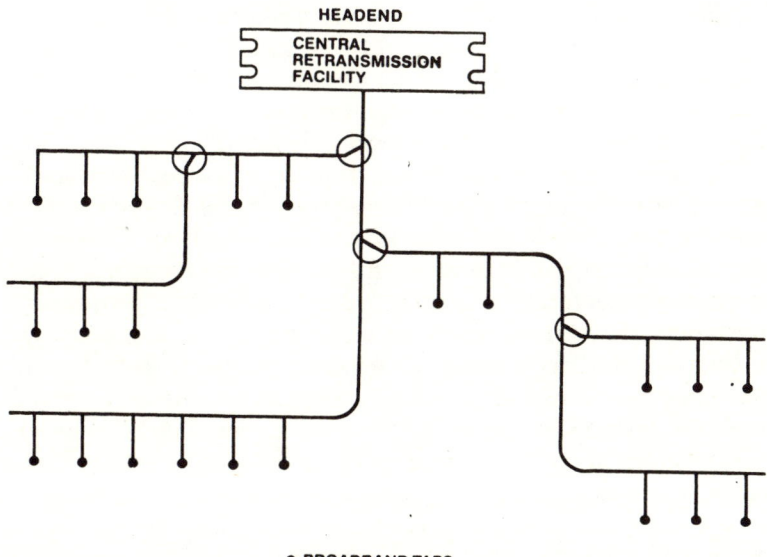

Fig. 3.8 Tree and branch cable topology.

into the cable at the headend and subscribers connect to the cable at the branches to receive the signals.

In a two-way cable system any signal injected into the cable system at a branch will always travel towards the headend. A two-way cable system is designed with components which give it this directional characteristic.

Signals injected in the forward direction into the cable system at the headend will percolate throughout the system and be accessible at all subscriber outlets. A simple analogy exists between the cold water distribution in a building and a cable system. If we imagine the headend to represent the cold water tank at the top of the building and the subscriber outlet to be the cold water taps on the floors below, then provided we have water in the header tank, water will flow when any of the taps are opened.

Signals injected in the reverse direction into the cable system at a subscriber outlet will always travel to the headend. A simple analogy exists here between the drainage system in a building and the cable system. If we imagine the subscriber outlets to represent washbasins, then allowing water to flow from washbasin into the drains is equivalent to injecting signals into the cable system at the subscriber outlets. Releasing water into the drainage system causes the water to flow towards the main sewers and nowhere else – this is equivalent to signals always travelling towards the headend. The drainage system is designed to be directional!

Using the fact that the cable system is designed to be directional, then selecting a frequency f_1, normally a high frequency, to be the forward frequency channel and frequency f_2, a low frequency, to be the reverse frequency, these two frequencies now define a full duplex data bus which exists throughout the cable system.

Modems may be connected to this data bus by tuning the modem to the correct frequency pair f_1, f_2. An example of this is shown in Fig. 3.9, where a master modem is located at the headend. The master modem transmits at frequency f_1 and receives at frequency f_2. All slave modems receive at frequency f_1 and transmit at frequency f_2. This configuration of modems is generally known as a *polled, multi-dropped system*. In this system all the slave modems share the data bus, and the master modem controls the slave modems by polling them over the forward channel. The master modem uses the forward channel to transmit control sequences or data to a selected slave modem. A particular control sequence to a selected slave modem would enable that slave modem to transmit data to the master modem at the headend. The control sequences and responses define a data communications protocol.

Using a data communications protocol and multi-dropped

modems enables two-way interactive communications to take place between a computer located at the headend and many terminal devices located elsewhere in the cable system.

Defining additional frequency pairs establishes new data buses on the same cable system. The bandwidth of the data channel will define the maximum data rate that can be supported.

In a modern data processing network processing may be located anywhere in the network and not concentrated at any central location. Therefore a network where communication takes place

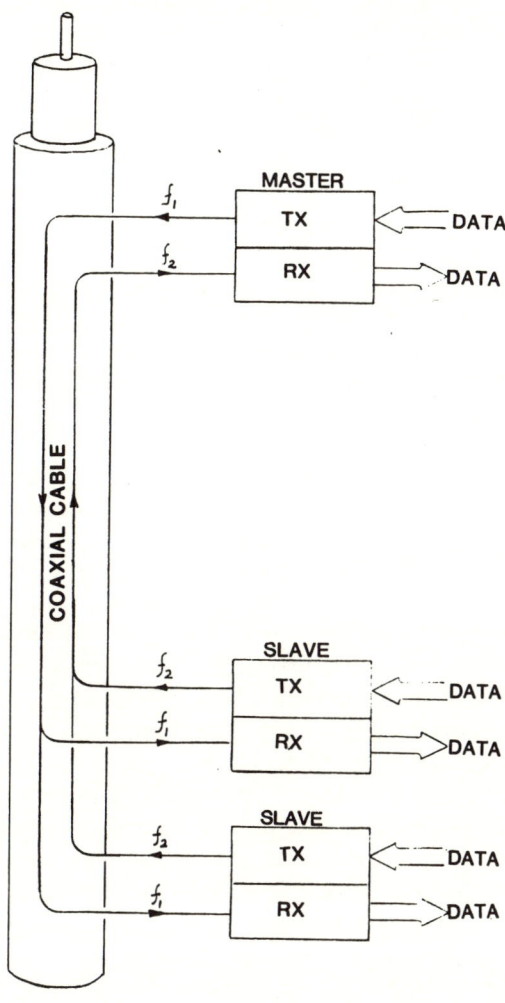

Fig. 3.9 Master/slave modems.

between a computer located at the headend and terminals located elsewhere in the network is inappropriate for many applications. What is required is peer-to-peer communication between processing nodes which may be located anywhere in the network. A network providing peer-to-peer communications can be constructed if we introduce a new component called a *frequency translator* at the headend.

The frequency translator receives a signal, in the reverse direction, from the cable system, usually at a low frequency f_{LOW}, amplifies and translates this signal to a high frequency, f_{HIGH}, and then broadcasts this signal in the forward direction into the cable system.

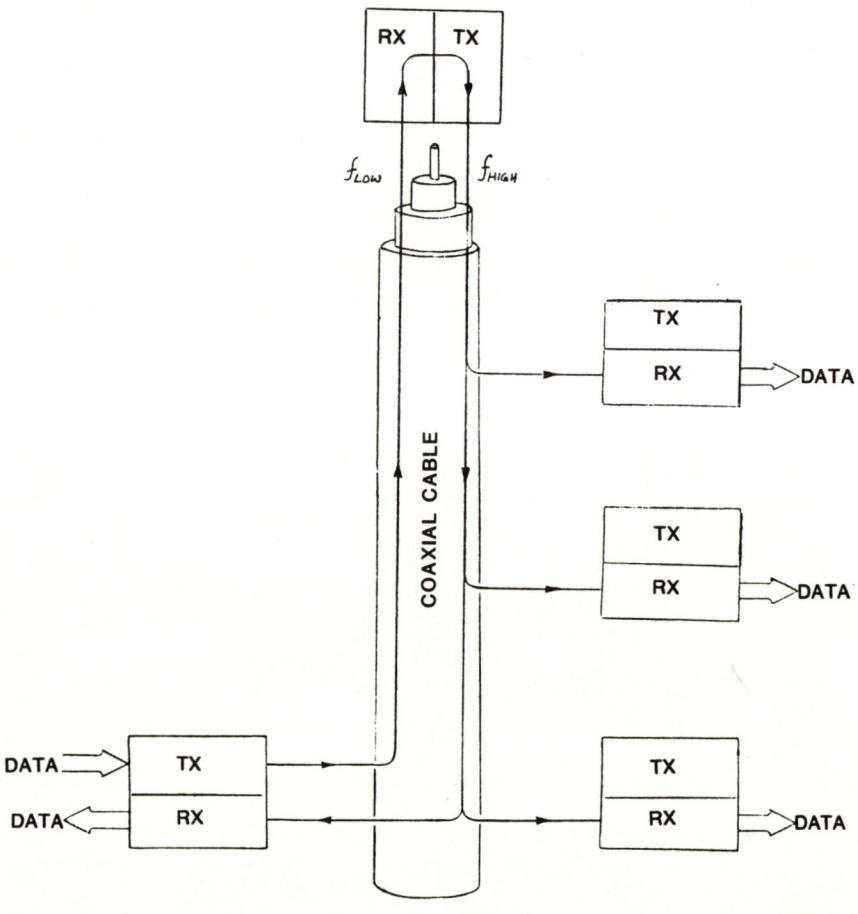

Fig. 3.10 Principle of broadcast bus systems.

Placing a frequency translator at the headend enables any transmitter located anywhere in the cable system to broadcast to all receivers tuned to the forward frequency, f_{HIGH} (Fig. 3.10). Obviously if two or more transmitters turn on simultaneously the signals would interfere or collide and the information being transmitted would be corrupted. A data communications protocol must therefore be established either to eliminate the possibility of collisions, or if collisions do occur to detect them and re-transmit the correct information.

The allocation of a pair of frequencies, f_1 and f_2, plus a frequency translator at the headend enables a data bus or information channel to exist throughout the cable system. Radio frequency modems tuned

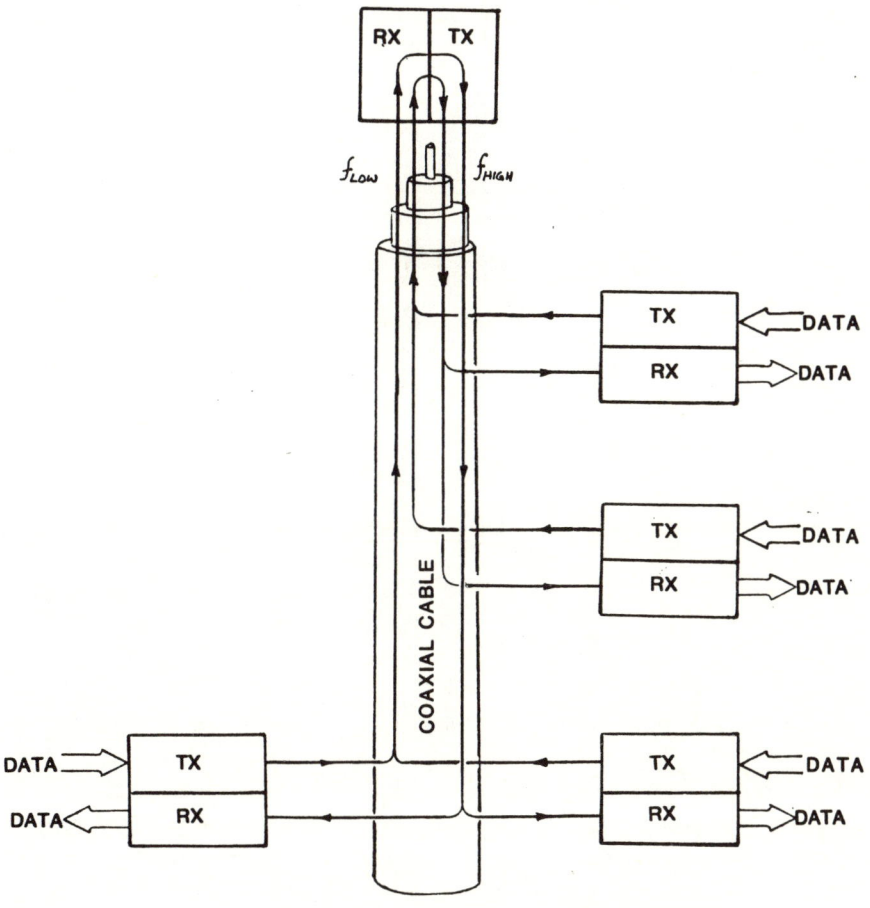

Fig. 3.11 Half-duplex private circuit.

to f_1 and f_2 allow access to this information channel. The addition of data communication protocols controls the access to the information channel so that collisions between transmitters are avoided or detected and recovered from.

An information channel defined by a pair of frequencies, f_1, f_2, and a frequency translator at the headend can be used as a single, half-duplex point-to-point circuit between any two locations on the cable system (see Fig. 3.11). The point-to-point communications is half-duplex because both locations cannot simultaneously transmit to each other; if they did the signals would interfere and the information transmitted would be corrupted.

Introducing a data communications protocol at a nominated

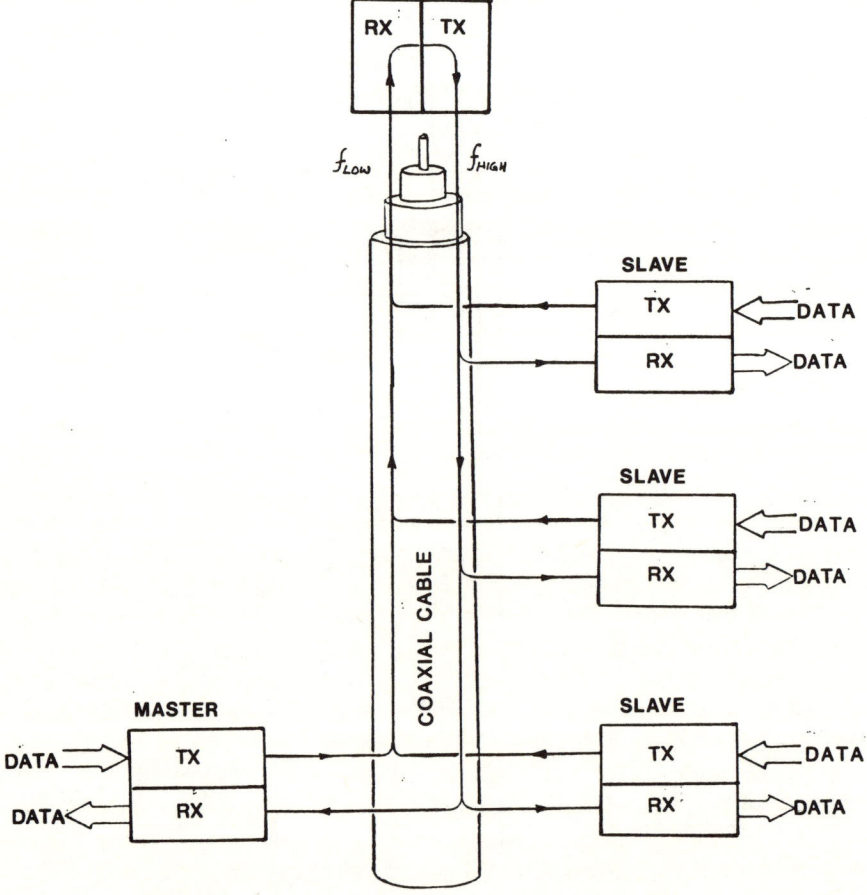

Fig. 3.12 Half-duplex multi-point/multi-dropped circuits.

location will make that modem the master. This location via a communications protocol can control communications between itself and all other slave modems tuned to the same information channel. This technique allows multipoint or multi-dropped circuits to be constructed between a master modem located in the cable system and a number of slave modems located anywhere in the cable system (see Fig. 3.12). The user has complete freedom to locate and move equipment anywhere within the geographic confines of the cable system without having to rewire. Many mainframe computer systems implement protocols that support multi-dropped communication circuits. These protocols and circuits can readily be supported via RF modems on a cable system.

With the rapid introduction of distributed processing via microcomputing systems and personal computers it becomes more and more imperative that data communications should take place on a peer-to-peer basis, rather than being restricted to communicating when instructed to, only between master and slave. New protocols have evolved to allow peer-to-peer communications to be established between processing nodes that share a common data bus or information channel. The two most relevant protocols that apply are:

- CSMA/CD. **Carrier Sense, Multiple Access, with Collision Detection.**
- Token bus.

CSMA/CD
CSMA/CD is a technique that anticipates conflicts and collisions amongst the users of the common information channel. The protocol acknowledges them as part of the design to allocate the common information channel bandwidth to users.

Carrier sense is the ability of each node to detect any traffic on the channel (listen-before-talking). Nodes will not transmit if they detect traffic on the channel. However, because of the time it takes for a signal to travel from a transmitting node to the headend and then to all receivers, two nodes could detect that the channel is free, since each will not yet have detected the signal of the other. In such a situation, a collision or interference of the two signals will occur.

The multiple access feature of CSMA/CD lets any node transmit a signal upon sensing that the information channel is free of traffic. In this way, a substantial portion of the waiting that is characteristic of polling or time division multiplexing techniques is eliminated. For instance, in polling, a node that is ready to transmit a message must wait for the completion of polling of all intervening nodes, whether or not they wish to transmit. CSMA/CD contains no such delays and therefore can allow quicker access and greater utilisation of the channel.

Collisions are detected by nodes while they are transmitting, by monitoring the received signal and comparing the characteristics of the received signal with the transmitted version (listen-while-talk). This comparison may be a simple bit-by-bit comparison between the transmitted message and its received image – if the comparison fails it is assumed that a collision has taken place. Non-complex characteristics of the transmitted and received signals may be compared and used as a criterion to reduce the probability of a collision being undetected. One such criterion is the comparison at the transmitting node of the transmitter's source address contained in the message. This source address is unique within the network and therefore a bit error in this field is interpreted as a collision. Other characteristics of the transmitted signal may also be monitored. For example, messages usually have a fixed preamble which is transmitted before the actual information portion of the message. If the end of this preamble is not detected at the correct time this violation may be used as an additional criterion to detect a collision.

Using CSMA/CD as a channel or media access protocol enables many nodes that produce bursty traffic to share the same channel. A big advantage of CSMA/CD as an access protocol is that it does not rely on any central control – all the control is distributed to the individual processing nodes. (See Fig. 3.13.)

TOKEN BUS
Token bus is an alternative method by which control is exercised between nodes wishing to access the same information channel. The token bus protocol allows the holder of a control token, which is a special packet of information, to have exclusive use of the transmission media without fear of interruption or collision from any other

---

Basic protocol:

(1)   When ready to transmit, first listen to channel (listen-before-talk). If channel is idle, begin transmission.

(3)   If channel is busy, wait until it becomes idle, then begin transmission.

(3)   During transmission, listen to channel and verify transmission (listen-while-talk).

(4)   If receiver does not match transmission, then collision occurred – multiple devices were transmitting. Wait for random (short) period and go to step 1.

Fig. 3.13 CSMA/CD.

node. The technique relies on the token being circulated from node to node.

In a broadcast network every node on the channel can hear all transmissions made by every other node. The token must therefore be explicitly addressed to the next node which is to have control of the channel. The nodes on the bus are arranged logically in a ring, although their physical locations may be quite different.

If a node wishes to transmit a message it will 'hold' the token and send its message specifying the destination address. If the holder of the token has no more messages to transmit it will put the token back into circulation, indicating that it has finished transmitting and giving other nodes the chance to gain access to the channel.

The token bus access protocol must include some means of placing the token into circulation when the network is first started as well as after any interruption in network operation. The circulation of the token also needs to take into account nodes that leave the network and others which join. A single node will usually be charged with this network management task.

Packet switched service

CSMA/CD and token bus protocols are channel access protocols used by network nodes to share efficiently the available channel bandwidths between the currently active nodes.

Packet switching or the routing of information packets to their correct destination via a broadcast bus is achieved as follows:

- The network node uses either CSMA/CD or token bus protocol to access the common channel.
- When the channel is available the node will broadcast an information packet to a destination node.
- The information packet will contain the address of both sender and destination nodes.
- Because the information packet is broadcast, all nodes sharing the same channel will receive the transmitted packet.
- All nodes compare the destination address contained in the received packet with their own node address.
- If the destination address matches the node address the node will retain the packet for further processing.
- If the destination address does not match the node address the packet will be discarded.

In a broadcast bus system the distributed intelligence contained in each node will select from all the packets being broadcast only those packets destined for the node, all other packets in the broadcast being ignored. The routing of packets is therefore achieved by a selection process at the destination node (see Fig. 3.14).

In a packet switching system, channel bandwidth is only used when a node is actually transmitting information. A packet switching system is therefore ideally suited to applications where the traffic from any particular node is intermittent and bursty in character. The channel bandwidth can therefore be shared between a number of communicating nodes.

Circuit switched service

Circuit switching is an alternative technique to packet switching. It

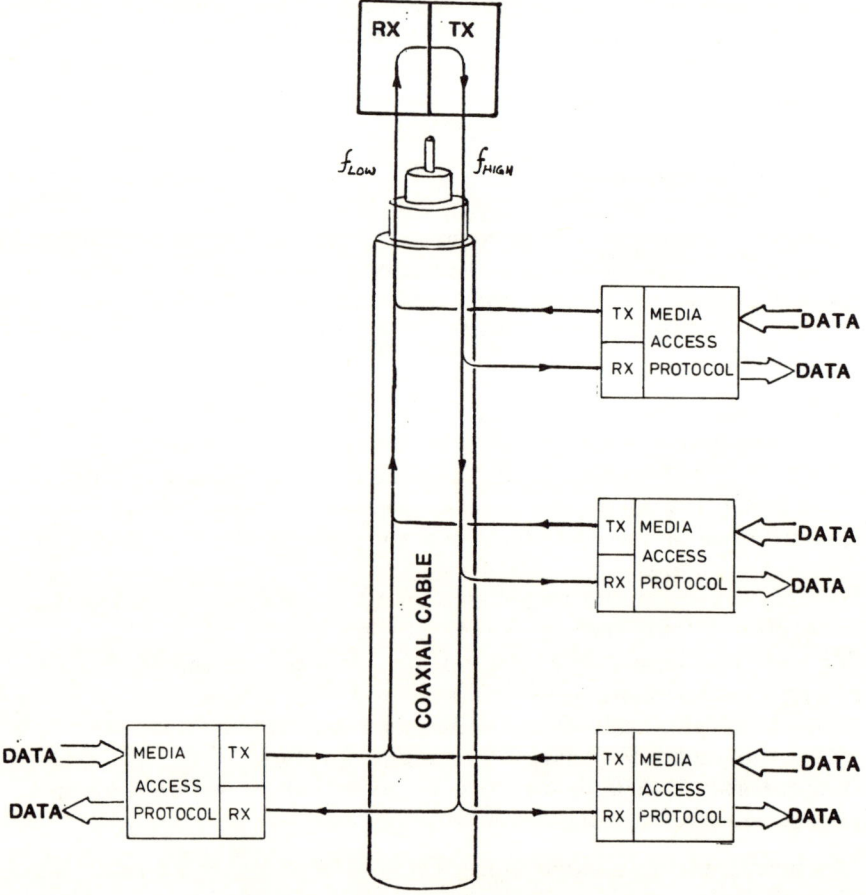

Fig. 3.14 Packet switched service.

provides a dedicated, full bandwidth channel between two communicating nodes, with the full channel bandwidth available to the two communicating nodes for as long as they are in communication with each other. See Fig. 3.15.

As explained earlier, a pair of carrier frequencies from the radio frequency spectrum of a cable system defines an information channel suitable for half-duplex communications – each node takes it in turn to communicate with the other node. Four carrier frequencies are required for full duplex communications to take place between nodes – this enables simultaneous transmission and reception of information to take place. The maximum rate of data transmission is determinded by the bandwidth of each channel.

By the use of frequency agile radio frequency (RF) modems, a pair of network nodes may be 'tuned' to operate at a particular channel. This pair of nodes has therefore been switched to a channel or circuit where the full bandwidth is available for use by the two nodes. Radio frequency modems, when not providing a circuit connection, will be connected to a common signalling channel. This channel is controlled by a network management processor.

The network management processor will accept requests from RF modems requesting a circuit to another RF modem. The RF modems will accept commands from the network management processor via

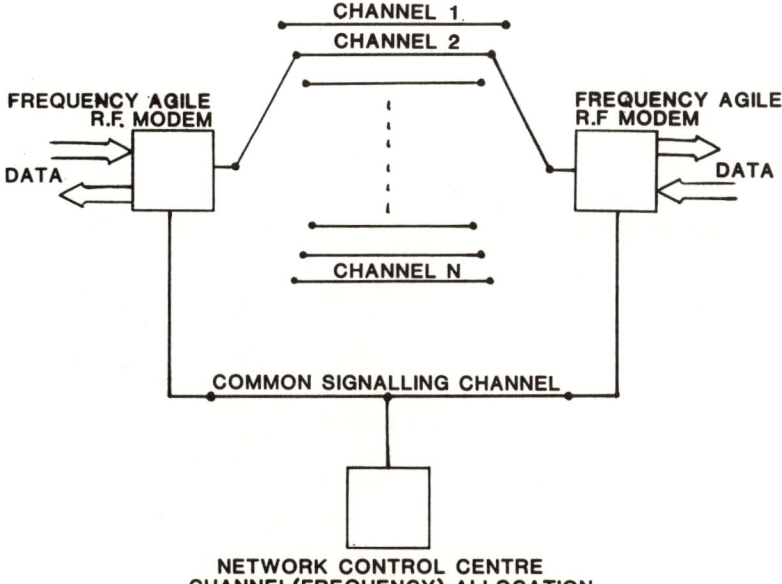

Fig. 3.15 Circuit switched service and/or private circuit.

the signalling channel to 'tune' its carrier frequencies to operate at an appropriate channel.

The network management processor knows the channels available on the cable to provide circuits between nodes. When a request for a circuit is received the network management processor will check if the destination node is busy (already connected to another node), and if the node is not busy the processor will command both RF modems to set their carrier frequencies to the same values and thereby allocate a channel not currently being used. If the destination node is busy the caller will wait and try again.

The number of channels allocated to the circuit switch will depend upon the number of nodes (subscribers) and the characteristics of the traffic between nodes. The characteristic of this traffic will be determined by the following parameters:

• Number of nodes.
• Number of calls per busy hour.
• Holding time.

Table 3.2

Bandwidth	Interface	Applications
30 – 50 khz	RS232	Low speed data < 19.2 kbps
300 khz	V.35, X21	Medium speed data, 64 kbps
6 – 8 MHz	CCITT G703	High speed data, 2.048 Mbps
6 – 8 MHz	Composite video	Video conferencing

Table 3.3

Information stream	Applications
Image	Broadcast TV
	Security TV
	Video conferencing.
Digital	Point-to-point data circuits
	Multipoint data circuits
	Packet switch service
	Circuit switch service.
Voice	Broadcast FM radio
	Telephony.

Implementing RF modems with different bandwidths and interfaces can provide circuit switching facilities for a variety of applications, as shown in Table 3.2.

MULTIPLE SUB-NETWORKS

From the previous sections it can be seen that a broadband coaxial cable distribution system will support multiple information streams. The information streams may be classified by the application or service provided, as shown in Table 3.3.

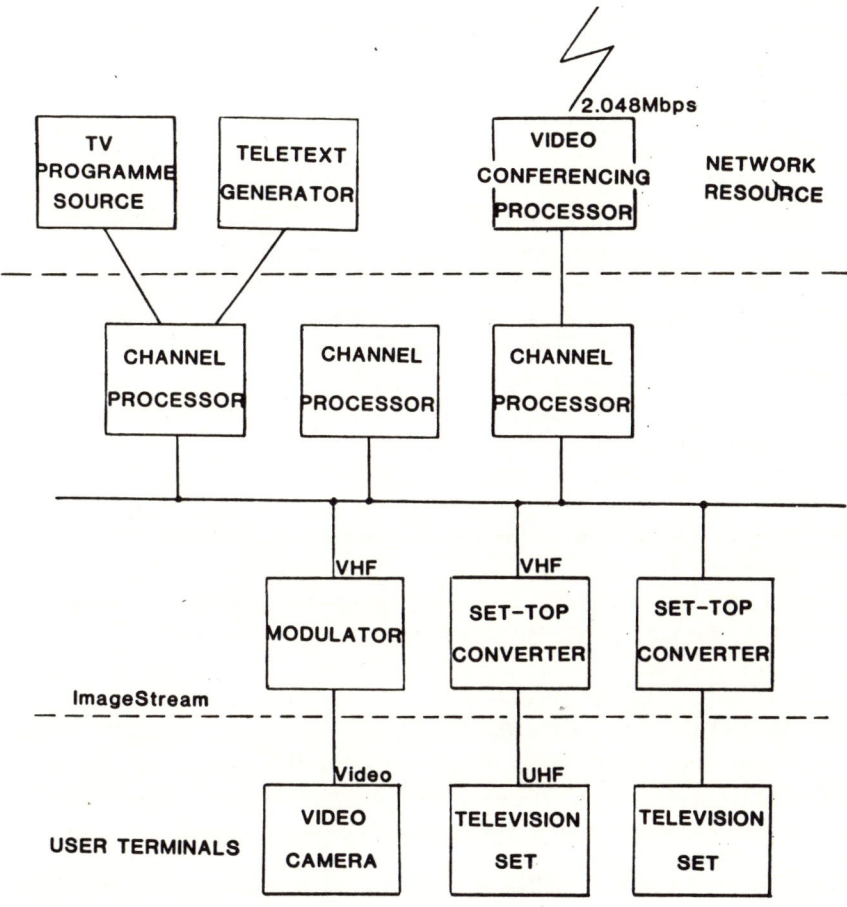

Fig. 3.16 Image services.

Image

Figure 3.16 illustrates three image services being provided to subscribers via a cable system:

- Broadcast TV.
- Teletext.
- Video conferencing.

BROADCAST TV

A TV programme source, e.g. video tape recorder, video disc or off-air broadcast, is modulated using a channel processor and broadcast to all subscribers on the cable. The TV programme is broadcast at very high frequency (VHF 100–400 MHz) via the cable system. At the subscriber's premises the TV programme is received by a set-top converter and the VHF TV signals are up converted to ultra high frequency (UHF 500 – 800 MHz) and input to a standard TV receiver. The set-top converter is a VHF tuner and gives the subscriber the ability to tune to broadcast material available on the cable system.

TELETEXT

Data can be carried in some of the unused TV line scan signals. This data will therefore be included and broadcast as part of the TV programme. A TV receiver with additional circuitry can decode this text and display it on the screen. A TV receiver with this facility is known as a teletext TV receiver. The text being continuously broadcast is formatted in pages suitable for display on the TV screen. The pages are numbered and a specific page may be selected by a user using a simple keypad.

Teletext is used by the off-air broadcast channel for a simple electronic information service. The amount of data is limited and the facility may be thought of as an electronic notice board. The same technique can be used via a cable TV network and each broadcast channel on the cable could have its own teletext notice board, containing pertinent information relevant to local events.

VIDEO CONFERENCING

Video conferencing is a technique whereby information contained in a television picture is compressed so that sufficient data may be transmitted via high speed telephone lines (1.544 – 2.048 Mbps) for the picture to be reconstructed at a remote location, with sufficient detail to enable groups of individuals at both locations to engage in a two-way conference. This technique allows groups of persons separated by large distances to be able to confer with each other and have visual contact without having to travel.

The electronic system that analyses the television picture and compresses the information is called a CODEC (coder/decoder). CODECs are expensive and this facility is either operated by a large corporation and shared by many divisions or departments, or offered as a service by a PTT organisation.

A video conferencing facility normally requires the groups to attend at video conferencing studios. If, however, the individuals within each group have access to an interactive two-way cable system, video images can be transmitted from any location on the cable system to the headend. A CODEC located at the headend would code up the images for transmission to the remote site. Any coded images received from the remote site would be decoded by the CODEC and the image would be available in a form suitable for broadcast across the cable system. With a CODEC located at the headend a video conference can be organised without requiring the group to attend at a central studio. Video conferencing can take place between remote offices that have access to cable systems with CODECs located at their headends.

This technique enables cable operators to offer a video conferencing service from city suburb to city suburb. A cable system therefore brings a video conferencing service to large potential markets.

Digital

As described above, the broadband cable system can support a variety of data communication services simultaneously on a single cable system. Each data service is allocated its dedicated frequency channel with sufficient bandwidth to support the bit rate of the digital information being carried.

The following data communication services are required by users and can be implemented to run on a broadband cable system:

- Transparent point-to-point or multipoint data circuits.
- Packet switched service.
- Circuit switched service.

TRANSPARENT DATA CIRCUITS
Transparent data circuits are implemented using RF modems. The RF modems use a dedicated frequency channel to provide point-to-point or multipoint circuits for asynchronous and sychronous bit streams of data (see Fig. 3.17).

Typical applications for transparent data circuits would be to provide circuits to connect terminal cluster controllers to mainframe computers. Mainframe computers implement their networking architecture using synchronous bit streams to implement a communications protocol between the mainframe and the terminal cluster controller. The terminals are connected locally to the cluster controller and cluster controllers are normally connected to the mainframe using a dedicated twisted pair of cables. Instead of providing dedicated cables, RF modems are tuned to dedicated frequency pairs on a single broadband cable to provide the necessary dedicated circuits. The broadband cable provides access to these dedicated frequency pairs from anywhere where access is available to the cable. If users are relocated, then provided access to the cable is available at the new location, computing services are restored to the users merely by relocating the terminals, cluster controller and RF modem and plugging into the cable. No rewiring is necessary.

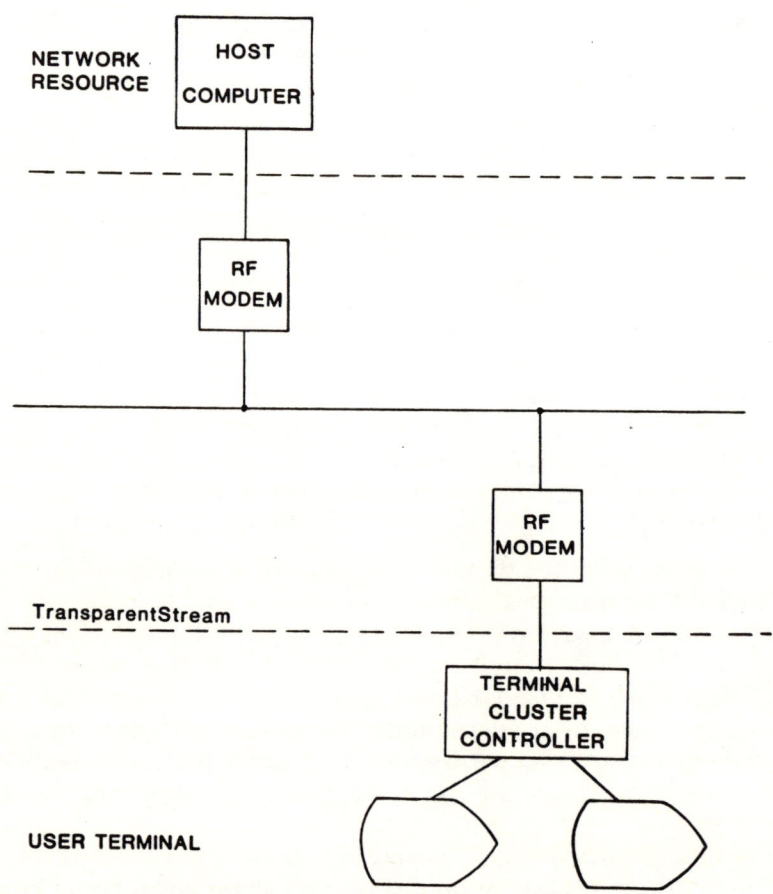

Fig. 3.17 Point-to-point data network.

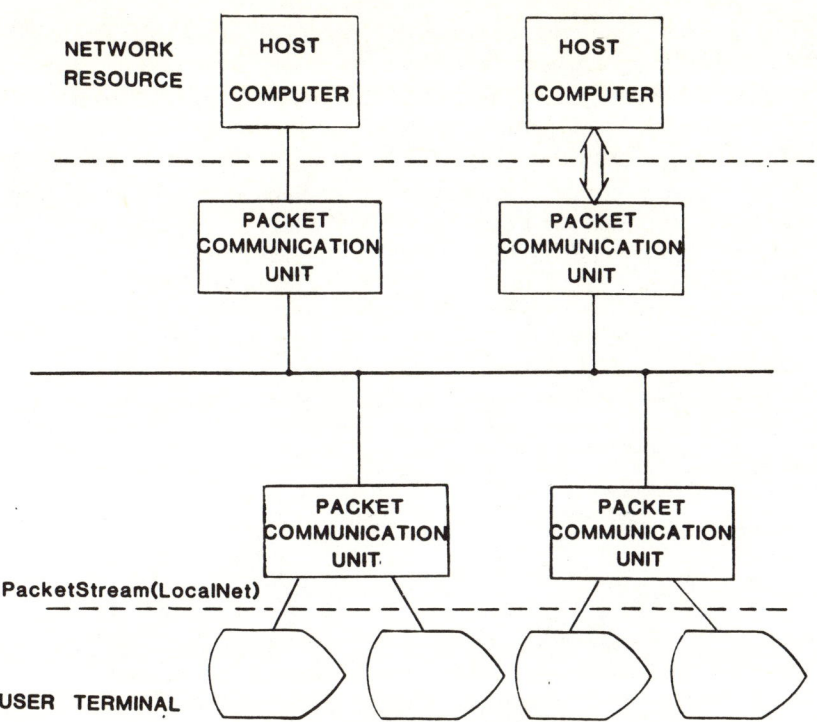

Fig. 3.18 Packet switched network.

PACKET SWITCHED SERVICE

Packet switching provides the data communication user with an enhanced communication service (see Fig. 3.18). Additional intelligence is now implemented within the network service to provide:

User initiated switching. Users may establish a path between a user's terminal and one of many service computers. This data circuit is user initiated and terminated by the user when not required.

Error control. The network will accept data from the user's terminal and deliver this data to the required destination with no errors. If any errors are introduced by the transmission paths of the network, the intelligence within the network will detect that an error has been introduced and will request a retransmission in order to correct the error.

Flow control. Communications speeds between packet nodes and user terminals operate at different data rates. Because packets are stored in each node there is no reason why the speeds should be the same. In particular, user terminals are generally of different speeds and the network operates for them as a speed changer. For example, a display unit operating at 1200 bps can communicate with a computer that has an access line to the network operating at 19.2 kbps.

The freedom to vary the flow of data carries a penalty, because the sender can create packets faster than the receiver is able to take them. This implies the need for flow control, with means that the receiver, in some way, controls the rate of the sender.

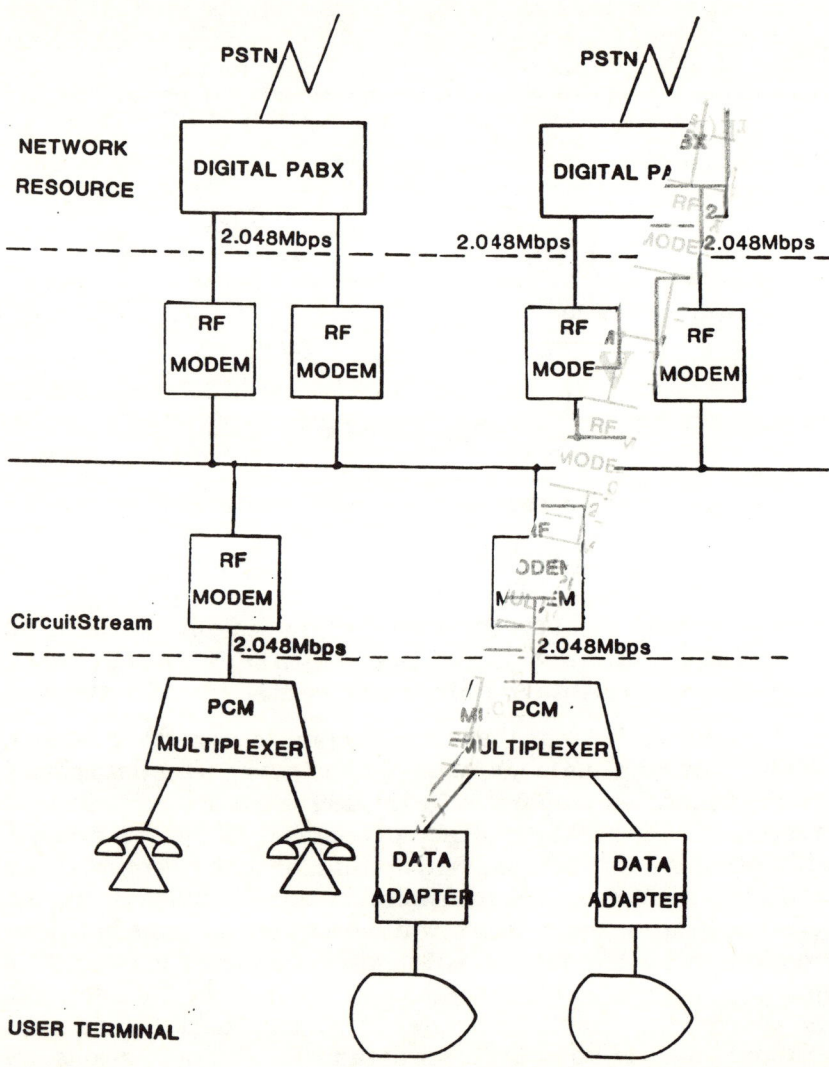

Fig. 3.19 Circuit switched network.

A packet communications unit (PCU) provides the intelligence within the network and performs packet assembly, disassembly, buffering, error and control. These services allow the setting up of packet switched connections between correspondent PCUs and sessions between their attached user devices.

CIRCUIT SWITCHED SERVICE

Using the high speed RF modems designed to support a data rate of 2.048 Mbps, the cable system can be used to provide PCM trunk circuits for digitally encoded voice or circuit switched 64 kbps data (see Fig. 3.19).

Digitally encoded voice generates 64 kbps of data. A single digital voice circuit therefore requires a transmission path that will support a data rate of 64 kbps. The European digital transmission hierarchy multiplexes thirty 64 kbps circuits on to a single 2.048 Mbps trunk.

A digital PABX will accept a 2.048 Mbps trunk circuit from remote PCM multiplexes and provide the circuit switching mechanism between individual 64 kbps circuits. Since all the data within such a system is now digital the digital PABX can switch both voice and data traffic.

It is currently not economic to use remote PCM multiplexes and

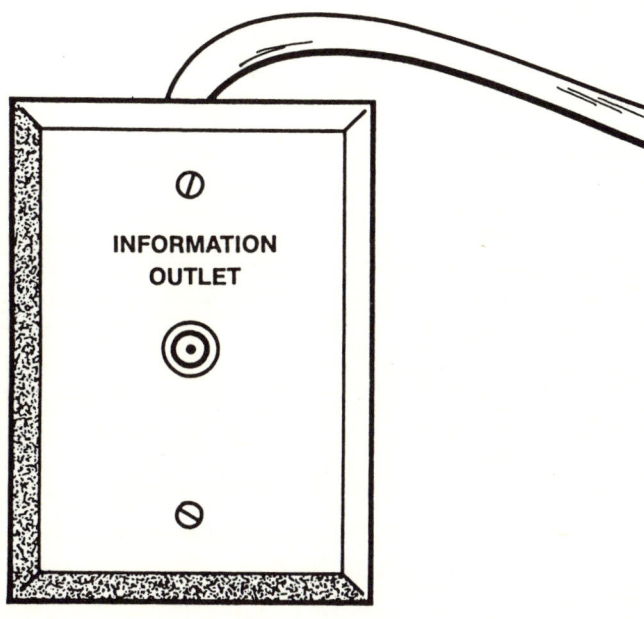

Fig. 3.20 User access point.

trunk all the voice circuits to a digital PABX for switching. The majority of telephone users still only use a conventional analogue phone providing a 'plain old telephone service' (POTS). This service is still economically provided using twisted pairs individually block wired to the PABX.

If the site has been wired with a broadband cable to carry data and video services then it may be both convenient and economic to provide 2.048 Mbps trunk circuits only for particular enhanced services such as circuit switched 64 kbps data.

The broadband cable could also be used to provide 2.048 Mbps trunk links between digital PABXs that are installed across the site.

Coaxial cable networking therefore supports many separate sub-networks, and all the sub-networks co-exist on a single cable distribution system. The users access the individual services available on the network via electronics that tune to the correct frequency channels and then decode the signals to derive the information being transmitted. The user access point is a cable TV outlet (see Fig. 3.20).

CHAPTER 4

Cable TV – An Historic Overview

In order to understand some of the properties of broadband data communications, it is useful to look at the historical development of the host technology, cable television, and its current usage. The problems of distributing large amounts of information over wide geographical areas are not new, and good engineering practice dictates consideration of past experience.

Cable television started in the USA in the late 1940s, primarily as a result of the low American population density. The existence of remote towns meant that a large section of the population were unable to receive any television signals at all, due to the distance or terrain between them and the transmitter. In Europe, such problems have usually been solved by the building of relay stations, but this solution was not practicable in the USA. First, there was no publicly financed body to build the relay stations, and the commercial gain to be made from the extra advertising revenue generated by building such a relay was negligible due to the low population density. (Outside the big cities, towns in America are small by European standards – 20 000 population is large.) Second, this low population density itself precluded the relay station approach, simply due to the number that would be required.

The initial demand was for any form of television service, and to this end CATV, or community antenna television, was started. In its basic form, CATV consisted of a very large aerial array connected to many television sets. The aerial was positioned for the best possible reception in that area. The connection to the television set was often made with open line twin feeder, a form of balanced radio frequency line not often seen today (see Fig. 4.1). This line is made of two parallel wires, often without insulation, spaced up to 5 cm apart, with wooden or ceramic spacers. The wires to the individual television sets were just hung onto the two conductors.

This form of RF distribution has many technical objections, and such systems must have provided extremely poor television pictures, with much 'snow' and 'ghosts'. At the time, however, any picture was regarded as better than no picture, and CATV operators were able to

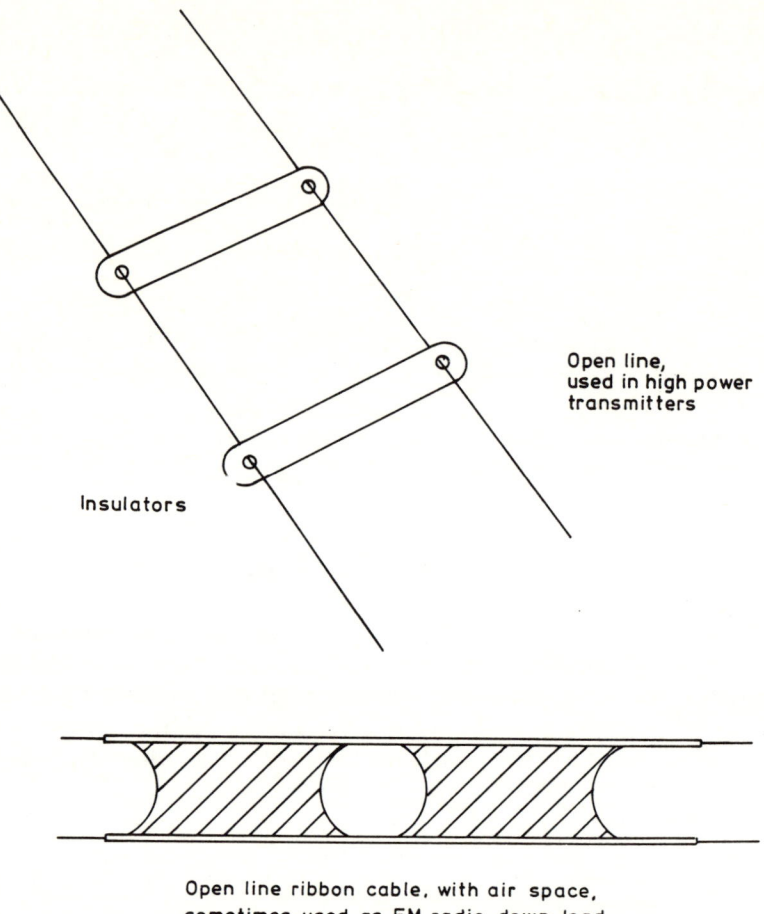

Open line,
used in high power
transmitters

Insulators

Open line ribbon cable, with air space,
sometimes used as FM radio down lead

Fig. 4.1 Types of parallel line twin feeder.

charge for connection to the system. As the number of television broadcast stations in the USA gradually increased, the CATV operators found it necessary to improve the simple systems, and nearly every part of the system was improved upon.

The first improvements consisted of adding amplifiers into the transmission system, to increase the signal level from the aerial array, and ensure that the subscriber at the end of the transmission line had enough signal at his television set. Along with this was the replacement of the open wire transmission lines with coaxial cable (see Fig. 4.2). This was probably the most important step. Theoretically, an open wire transmission line is every bit as good as a coaxial

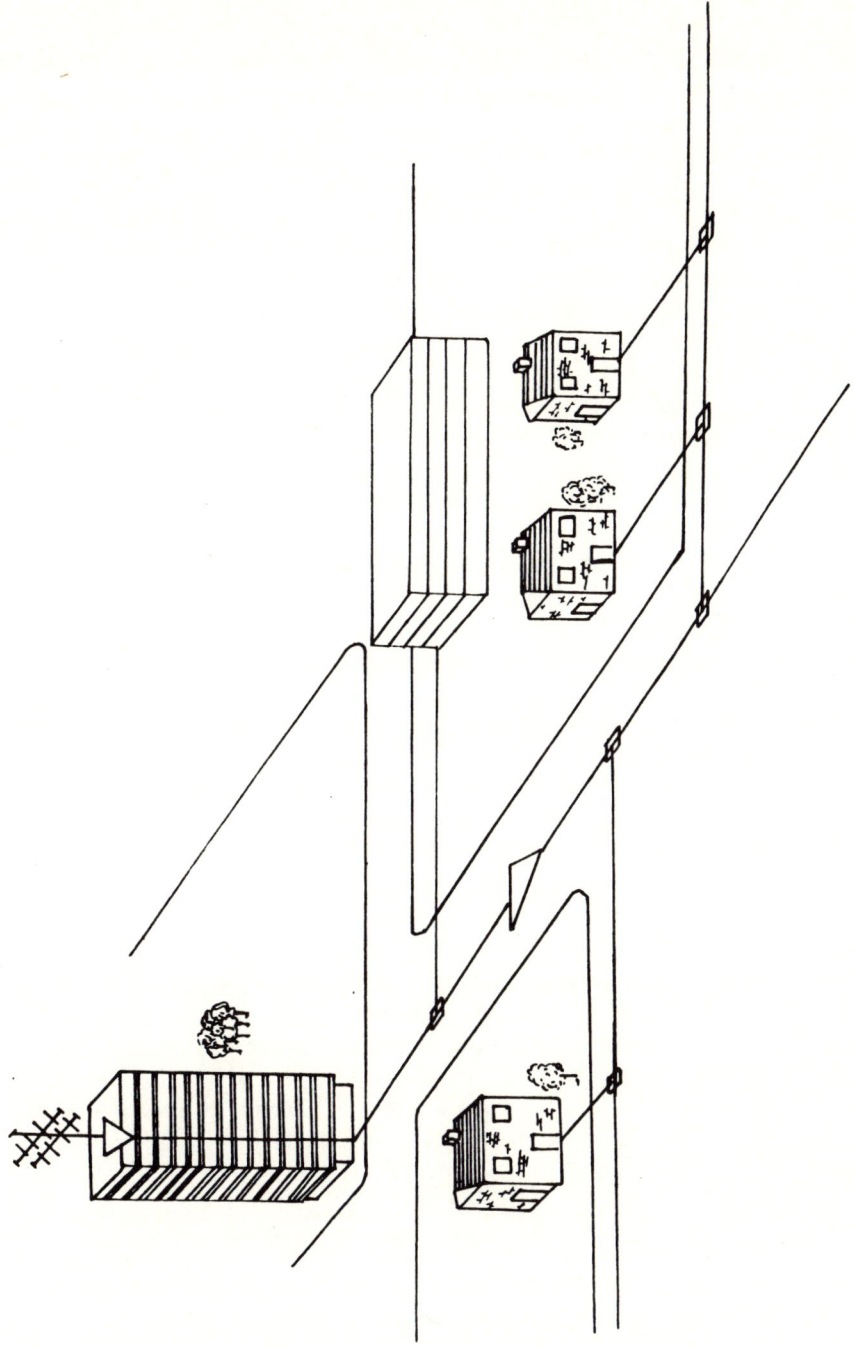

Fig. 4.2 A CATV system.

cable at transmitting television or radio signals along its length. It is, however, a 'balanced' system, where the electromagnetic fields around each conductor 'balance' each other, and no energy is lost to the outside world. Similarly, energy from the environment should not affect a balanced line. In practice, the balance is impossible to achieve, simply due to factors such as minute deviations in spacing, or insulator size. The result was that such large, open line systems picked up a large amount of noise from nearby electric motors, etc., whilst at the same time radiating the signals sent down them.

Coaxial cables, whilst more expensive, are known as 'unbalanced' systems, in that the outer conductor or screen is earthed, and cannot either radiate or pick up noise. (Again, in practice, this happens to a small extent, but the system is much better than an open line system.) Open line transmission systems are almost unused today, the few remaining applications being in high power radio transmitters, where coaxial cables would burn up, and for cheap FM aerial downleads in the USA.

The combination of amplifiers and coaxial cable ensured that the subscribers to a CATV system got enough signal, and not much noise, so that the 'snow' problem was overcome. The increasing number of subscribers on a cable system brought new problems, not connected with level, but with matching. To get an idea of a matching problem, we can use the analogy of a water pipe. In a water pipe, every new tap that we add along its length increases the flow requirement. This is easily dealt with by raising (within limits) the water pressure. In the same way, new subscribers to a CATV system can be dealt with by raising the signal level or signal voltage, by using an amplifier. Every tap that we add, however, introduces a slight roughness in the pipe wall, which causes the previously smooth flow to become turbulent, and if the junctions are very rough, we can end up by seriously restricting the flow. Connecting in a 'drop' or 'tap' coaxial cable to a subscriber is exactly like this. If the connection is not made properly, there will be a great deal of electromagnetic 'turbulence', causing the signals to be partially reflected and distorted. The result is that a coaxial cable with poorly made 'tap' connections will only pass distorted signals, leading to fuzzy edges and 'ghost' images on the television screen.

The first subscriber taps were of the very worst variety – simply pins stuck into the cable, after a small section of the outer screen had been removed. In many ways these removed most of the advantages of coaxial cable: they allowed the system to radiate, and they were not matched to the 'flow rate' – the 'impedance' of the cable. The first advance in the tap area was the so-called *resistive tap*. This made a much better match to the coaxial cable, and had the advantage that only a proportion of the power on the cable was fed to the television receiver. For the first time, the signal level at each subscriber could be accurately adjusted, simply by changing the amount of power tapped off in the resistive tap. Subscribers near the start of the system, or headend, where the main cable level was high, would now receive the same level at the television set as a subscriber at the end of the system. As the taps were properly matched to the cable, there was no longer a 'ghost' problem.

Such systems work quite well when all the subscribers behave rationally, but unfortunately, there is always someone who wishes to upset things. A resistive tap system is very vulnerable to anyone wishing to inject signals onto the cable, as exactly the same tap can be used to send signals along the cable in both directions. Similarly, a pin stuck in the drop lead will short out the signals on the cable.

Since the CATV operators depended on the provision of uninterrupted television service for their income, the state of affairs where

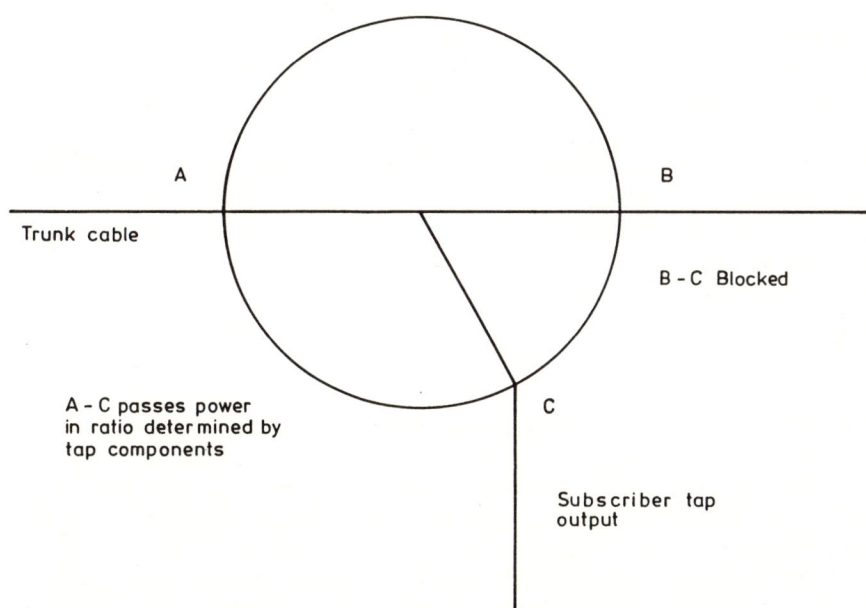

A subscriber connected to C can only receive signals coming from A. Similarly, any signal injected at C can only pass through the tap to A, and hence towards the headend of the CATV system. The tap actually works as a power splitter, splitting the available signal between the trunk cable and the tap output.

Fig. 4.3 The directional tap (transformer coupled).

the operator was at the mercy of the subscribers could not be allowed to continue. The solution to the problem was the directional tap (Fig. 4.3) where, due to the use of transformer coupling, only signals from the headend of the system can be received. Signals from the other direction, further down the system, are blocked. A subscriber can still inject a spurious signal onto the cable, but it will only be able to travel towards the headend. All the taps between the injection location and the headend will block this signal from entering their drop leads, and the spurious signal will end up at the headend itself, where it can be dealt with by use of filters. In much the same way, a short circuit on a drop lead will cause loss of signal on that drop only, as the transformers in the tap still provide a good match to the main cable system, because there is no direct connection between subscriber and cable.

The main components of cable television are thus the cable, amplifier and directional tap. These three components form the backbone of all CATV systems. Recent history has seen the standardisation of specifications for CATV services, to give high quality pictures without undesirable effects such as radiation from the cable. Such industry specifications enable the mixing of differing makes of CATV equipment on a single system. Currently, CATV services in the USA are offering an ever expanding number of television channels, as well as interactive videotex systems. Scrambling of the TV picture has become standard to prevent theft of service, and new devices such as addressable taps, controlled by a computer in the headend, allow implementation of functions such as electronic 'locking' of the tuning control.

CHAPTER 5

Spectrum Allocations

Frequency spectrum is the amount of 'space' taken up by a radio transmitter, or, in a cable system, the amount of 'space' on that cable – a measure of the capacity of the system.

Frequency is measured in hertz, a unit of one cycle per second. As most radio transmissions take place at very high frequencies, it is common to use units of millions of hertz or megahertz (MHz). The most common use of the megahertz unit is on a VHF radio dial, which usually covers from around 88 MHz to 108 MHz – thus having a 20 MHz wide frequency spectrum. In this space there is easily room for some 50 or more VHF radio stations, although in the UK the numbers are normally very much less. These radio stations have unique frequencies, and do not interfere with each other. In exactly the same way, signals on a broadband cable system have their own frequencies, unique to them, and do not interfere with each other.

In the case of CATV distribution, all the signals go in the same direction – a forward only system. The television signals are fed into the system at the headend, and are then distributed to the subscribers, each TV channel on a different frequency. Though the standards vary a little from country to country, a forward only system usually starts around 47 MHz and provides a continuous frequency space up to 300 MHz. On newer systems, designed to cater for more TV channels, the frequency spectrum is extended to 440 MHz.

Typical television capacity (American television channels, which take up only ¾ of the space of UK television channels) are as follows:

47 – 300 MHz: 35 channels
47 – 330 MHz: 40 channels
47 – 360 MHz: 46 channels
47 – 400 MHz: 53 channels
47 – 440 MHz: 60 channels

Typically, such a fully loaded forward only system could carry about 40 UK channels.

The ability to carry data on a cable system was a direct result of the search of American CATV operators for new types of service to sell to their customers. The operators were not interested in providing data services, but were looking for a means of providing a 'tiered' service structure. A basic CATV service provided all the channels on the cable to every subscriber for a simple monthly rental charge. The

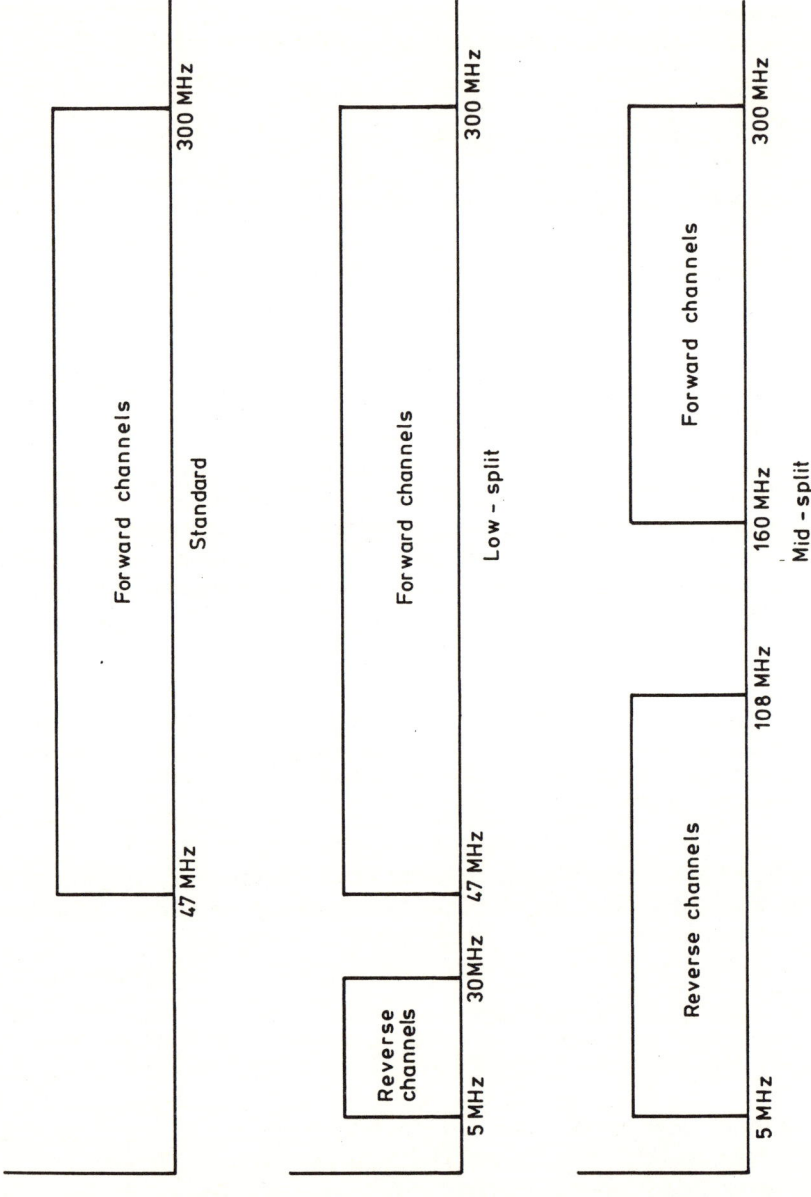

Fig. 5.1 Standard, low- and mid-split cable spectrums.

operators, seeking to maximise the profitability of the cable system, wanted to be able to charge extra for specific programmes. This could be done with a meter on the customer's premises, recording which station was watched, and for how long, but such meters are unavoidably prone to customer interference, costly to read, and inflexible with regard to tariff changes. The CATV operators thus wanted a means of using the existing transmission system in the other direction and the sub- or low-split system was created (see Fig. 5.1). In a sub-split system, the forward channels remain exactly as for a forward only system, but a small reverse band is added between 5 and 30 MHz (again, there are small variations in the exact frequencies around Europe). If there seems anything strange in a cable carrying signals in two directions at the same time, it should be remembered that the air around us carries signals in all possible directions from many hundreds of transmitters. As long as the frequencies are different, no interference will occur. The small reverse band of frequencies is used to provide such services as 'pay per view', where whenever the television set, or set-top decoder in a scrambled system is tuned to a specific channel a data transmission takes place from the set to the headend billing computer, using a radio channel in the reverse spectrum (see Figs 5.2 and 5.3). (This form of CATV will not be permitted in the UK, due to the invasion of privacy implications.)

Most cable systems could be converted from forward only systems to sub-split, simply by adding reverse band amplifiers.

Another function of sub-split systems is to transmit locally originated television pictures to the headend. An almost uniquely American requirement is for televising of events such as meetings of state school governors, church events, etc., on a regular basis. Many of the multi-channel CATV systems allocate one or more spare channels exclusively for such usage. A small studio will thus be built in the town hall, church office or wherever, and the programmes originated there will be sent along the cable system to the headend, where the frequency will be changed to one in the forward path, and then sent out on the whole network. This form of frequency conversion is a result of the directionality of the taps.

True data systems will have just as much data going in one direction as in another, so for data systems, and specialised TV applications, the mid-split system was created (see Fig. 5.1). A mid-split system has a reverse spectrum from 5 to 108 MHz and a forward spectrum from 160 to 440 MHz (with the latest amplifiers). Although there is still less reverse bandwidth, the imbalance is not so severe. The large gap between the reverse and forward spectrum parts is to ensure that only the correct frequencies get amplified by the forward or reverse amplifiers respectively.

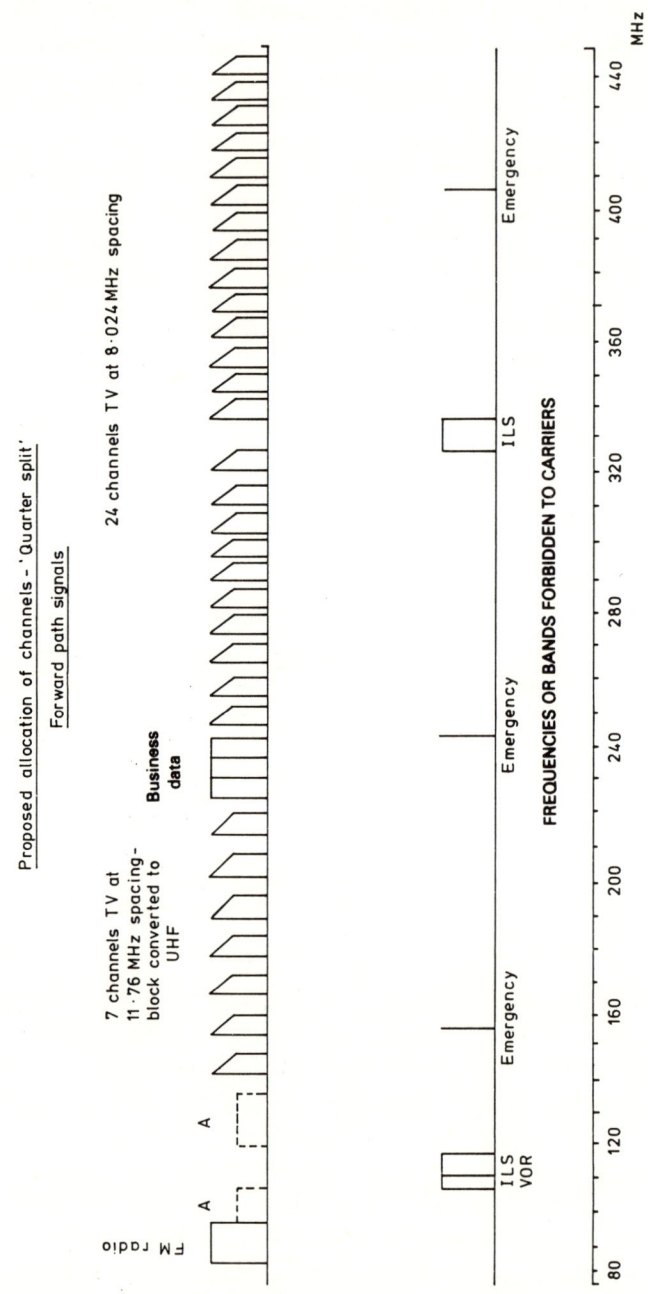

Fig. 5.2 Thorn EMI System Spectrum.

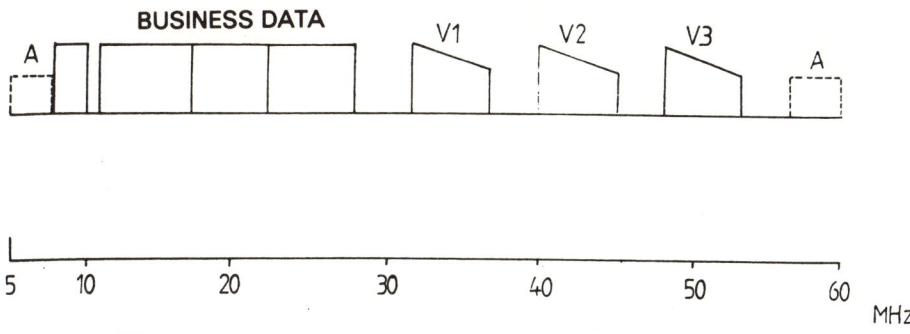

Fig. 5.3 Cable TV: allocation of channels (return path).

When a very large amount of bandwidth is required, a dual cable system may be installed. A dual cable system is simply two forward only systems, side by side, one cable carrying forward channels and the other carrying reverse channels. Thus the complete spectrum of the cable is available for both forward and reverse signals. A dual cable system is marginally easier to set up than a sub- or mid-split system, but the installation costs will be higher.

There are various mixtures of dual cables in CATV in the USA. Some systems use two cables, but both operating in the mid-split configuration, to give some 90 television channels, and a large return path. Obviously, all sorts of system configurations are possible.

In data networks, the best known dual cable system is WangNet, where the entire spectrum of one cable is used as the forward channels, and the entire spectrum of the other as the reverse. Wang have departed from standard CATV practice, and have built their own amplifiers to cover the range 10–350 MHz, a non-standard forward frequency range.

Most of the available broadband data equipment is designed for mid- or sub-split cable systems, but in general can be easily modified for dual cable operation.

TV CHANNELS AND DATA CHANNELS

Because broadband data systems are based on American standard cable television technology, it is common to refer to frequencies on the cable in terms of the standard USA channel designations for television. This can be a source of great confusion, especially when the various European channel designations are also used.

Basically, a channel in radio frequency terms refers to an unspecified amount of frequency space or 'bandwidth'. The amount of bandwidth used for any particular application is a function of the information rate, and the transmission quality offered by the transmission media. This interrelation of media quality and information rate is well known. It is sometimes necessary to repeat things over the telephone, when the line quality is poor – the information rate being therefore reduced. On a poor quality transmission system, there are some technical ways around this, by using different modulation techniques, but all these techniques rely upon increasing the bandwidth – a form of repetition of the information, but in frequency, not time.

The channel width thus depends on the quality of the transmission system, and the modulation technique used. Cable systems are quite good quality, and can transmit television pictures in just the same amount of frequency space as there is information in the picture. For an American picture, this is around 5 MHz, and nearly 8 MHz for a British picture. How these figures are arrived at is shown in Fig. 5.4.

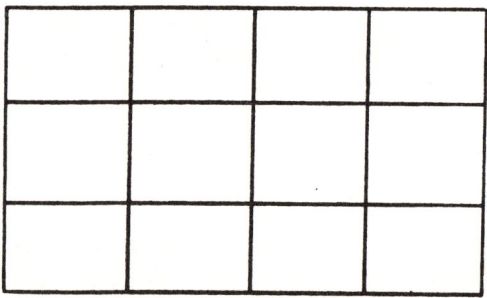

Each square represents a picture element. The number of picture elements on any line on a TV system is equal to the number of half-cycles of maximum video frequency. If the aspect ratio of the picture, 'A', is equal to picture width/picture height, the number of picture elements will be $L \times L \times A$, where L is the number of lines.

In practice, when horizontal or near horizontal bars are displayed, lines are 'broken up', and the number of picture elements reduced. This factor is known as the 'Kel factor' and is typically 0.7. Thus the actual number of picture elements is $L \times L \times A \times K$. If each element is a half-cycle, then the number of cycles is $(L \times L \times A \times K)/2$. Taking the UK TV picture, 625 lines with an aspect ratio of about 1.5, the number of cycles in a complete frame is $(625 \times 625 \times 1.5 \times 0.7)/2 = 205\,000$, and this frame occurs every 1/25 of a second, giving 5 226 953 cycles per second or 5.23 MHz.

Fig. 5.4 Information content of a TV picture.

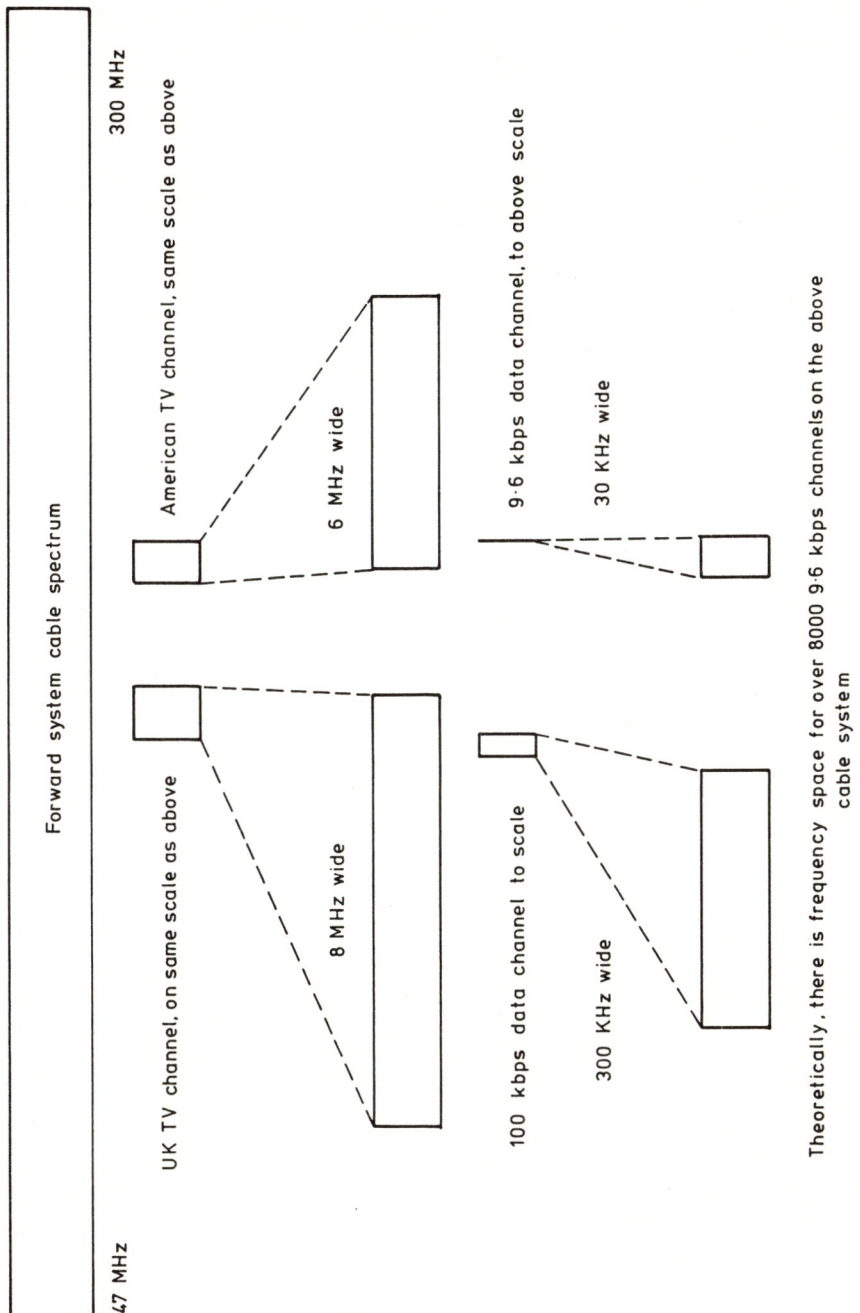

Fig. 5.5 TV channels and data channels.

BAND I 405 line

| Channel | Carrier frequencies (MHz) | |
	Vision	Sound
1	45.0	41.5
2	51.75	48.25
3	56.75	53.25
4	61.75	58.25
5	66.75	63.25

BAND III 405 line

| Channel | Carrier frequencies (MHz) | |
	Vision	Sound
6	179.75	176.25
7	184.75	181.25
8	189.75	186.25
9	194.75	191.25
10	199.75	196.25
11	204.75	201.25
12	209.75	206.25
13	214.75	211.25

BAND IV 625 line

| Channel | Carrier frequencies (MHz) | |
	Vision	Sound
21	471.25	477.25
22	479.25	485.25
23	487.25	493.25
24	495.25	501.25
25	503.25	509.25
26	511.25	517.25
Group A 27	519.25	525.25
28	527.25	533.25
29	535.25	541.25
30	543.25	549.25
31	551.25	557.25
32	559.25	565.25
33	567.25	573.25
34	575.25	581.25

BAND V 625 line

| Channel | Carrier frequencies (Mhz) | |
	Vision	Sound
39	615.25	621.25
40	623.25	629.25
41	631.25	637.25
42	639.25	645.25
43	647.25	653.25
44	655.25	661.25
Group B 45	663.25	669.25
46	671.25	677.25
47	679.25	685.25
48	687.25	693.25
49	695.25	701.25
50	703.25	709.25
51	711.25	717.25
52	719.25	725.25
53	727.25	733.25
54	735.25	741.25
55	743.25	749.25
56	751.25	757.25
57	759.25	765.25
Group C/D 58	767.25	773.25
59	775.25	781.25
60	783.25	789.25
61	791.25	797.25
62	799.25	805.25
63	807.25	813.25
64	815.25	821.25
65	823.25	829.25
66	831.25	837.25
67	839.25	845.25
68	847.25	853.25

VHF/625 LINE CHANNEL ALLOCATION

BAND I

| Channel | Carrier frequencies (MHz) | |
	Vision	Sound
A	45.75	51.75
B	53.75	59.75
C	61.75	67.75

BAND III

| Channel | Carrier frequencies (MHz) | |
	Vision	Sound
D	175.25	181.25
E	183.25	189.25
F	191.25	197.25
G	199.25	205.25
H	207.25	213.25
I	215.25	221.25

A television picture, however, is tolerant of some loss of information - the human brain will supply a large amount of missing information. Data, in contrast, can tolerate very few errors indeed, so the bandwidth taken up by a data signal is usually between 2 and 3 times the data information rate. Thus for a typical 9.6 kbps data stream, the cable data channel bandwidth will be around 30 kHz, and for 100 kbps, the bandwidth will be about 300 kHz.

It should be noted that any division of the available cable frequency spectrum into channels, of whatever width, is completely arbitrary, and that the various channel designations are merely mnemonic aids. The important parameter in any broadband system is the bandwidth of the information source, and how closely such signals can be spaced.

The spacing of television or data signals (see Fig. 5.5) from each other is normally rather more than the exact signal bandwidth, to take account of imperfect receiver and transmitter design.

Figure 5.6 shows frequency allocation for UK TV channels.

CHAPTER 6

Measure of Signal Levels

THE dB

In radio frequency systems, large differences can occur between, for instance, a transmitter output level, and a receiver input voltage. If direct units such as volts or watts are used to express these numbers, confusion can arise due to the very wide ranges (up to 10^9). Similarly, when dealing with noise or signal voltages in an amplifier chain, the effects are always multiplicative, not additive. The system level calculations become tedious and prone to error.

A basic property of logarithms is that multiplication becomes simple addition, so the dB system was devised to exploit this property in radio system calculations. There are, however, too many different systems in use, and this causes complications.

First, the dB:

The dB is a ratio – it *never* expresses a definite quantity.

Its prime use is in power ratios, where the dB is defined as:

dB = 10 log power 1/power 2

As power is proportional to voltage squared, the dB can be defined as:

dB = 20 log voltage 1/voltage 2

Hence, a gain of 20 dB means a power gain of 100 times and a voltage gain of 10 times. (See fig 6.1).

dB	Ratio	dB	Ratio	dB	Ratio
0	1.00	20.0	10.0	40.0	100
0.5	1.06	20.5	10.6	40.5	106
1.0	1.12	21.0	11.2	41.0	112
1.5	1.19	21.5	11.9	41.5	119
2.0	1.26	22.0	12.6	42.0	126
2.5	1.33	22.5	13.3	42.5	133
3.0	1.41	23.0	14.1	43.0	141
3.5	1.50	23.5	15.0	43.5	150
4.0	1.59	24.0	15.9	44.0	159
4.5	1.68	24.5	16.8	44.5	168
5.0	1.78	25.0	17.8	45.0	178
5.5	1.88	25.5	18.8	45.5	188
6.0	2.00	26.0	20.0	46.0	200
6.5	2.11	26.5	21.1	46.5	211
7.0	2.24	27.0	22.4	47.0	224
7.5	2.37	27.5	23.7	47.5	237
8.0	2.51	28.0	25.1	48.0	251
8.5	2.66	28.5	26.6	48.5	266
9.0	2.82	29.0	28.2	49.0	282
9.5	2.99	29.5	29.9	49.5	299
10.0	3.16	30.0	31.6	50.0	316
10.5	3.35	30.5	33.5	50.5	335
11.0	3.55	31.0	35.5	51.0	355
11.5	3.76	31.5	37.6	51.5	376
12.0	3.98	32.0	39.8	52.0	398
12.5	4.22	32.5	42.2	52.5	422
13.0	4.47	33.0	44.7	53.0	447
13.5	4.73	33.5	47.3	53.5	473
14.0	5.01	43.0	50.1	54.0	501
14.5	5.31	34.5	53.1	54.5	531
15.0	5.62	35.0	56.2	55.0	562
15.5	5.96	35.5	59.6	55.5	596
16.0	6.31	36.0	63.1	56.0	631
16.5	6.68	36.5	66.8	56.6	668
17.0	7.08	37.0	70.8	57.0	708
17.5	7.50	37.5	75.0	57.5	750
18.0	7.94	38.0	79.4	58.0	794
18.5	8.41	38.5	84.1	58.5	841
19.0	8.91	39.0	89.1	59.0	891
19.5	9.44	39.5	94.4	59.5	944
				60.0	1000

20 dB = ratio of 10(10^1) 80 dB = ratio of 10 000 (10^4)
40 dB = ratio of 100 (10^2) 100 dB = ratio of 100 000 (10^5)
60 dB = ratio of 1000 (10^3) 120 dB = ratio of 1 000 000 (10^6)

Fig. 6.1 dB to voltage ratio.

THE dBmV

Unlike the dB, the dBmV is not a ratio, but an absolute level quoted as a ratio to a fixed point. That fixed point is 0dBmV. By definition:

0 dBmV = 1 mV across a 75 Ω resistor

Thus a level of 20 dBmV is 20dB higher in voltage across the 75 Ω resistor (or cable) than 0dBmV. As 20 dB is a voltage ratio of 10:1, 20 dBmV is in fact a voltage of 10 mV across the 75 Ω load. (See Fig. 6.2.)

Two other dBs can be found: dBm and dBμV. By definition:

0dBm = 1 mW across 50 Ω
(often found in signal generators)
0 dBμV = 1 microvolt across 75 Ω
Note: 0 dBmV = 60 dBμV

The dBμV is often found in community antenna TV distribution systems.

SIGNAL ATTENUATION

Radio frequency signals on a coaxial cable system are subject to attenuation, because of the effective high frequency resistance of the components. The basic causes of attenuation are the capacitance to earth, which tends to short circuit high frequencies, and the inductance through the cable system components, which tends to block the higher frequencies. The overall effect is that cable attenuates more strongly the higher the frequency.

Passive components like taps and splitters are designed to have as near constant attenuation at all system frequencies as possible. The amount of attenuation in a system component is called its *insertion loss*, and for passive components with constant loss this is known as *flat loss*.

Conventionally, loss is measured in dB, and cable specifications will give the loss per unit length (100 m) at differing frequencies. A system must be designed so that the amount of loss from the start or headend to any output port is controlled.

When it is necessary to insert extra attenuation, this is done with a device known as a *pad*, which is a resistive network, designed to be frequency independent, with a specific attenuation value. Typically, pads are available in 3dB steps from 3 to 30dB. For bench use, switched attenuators are available, where the required attenuation (in dB) can be set up on rotary switches. Pads are used, rather than just a resistor, as they offer a good match to the cable system and do not introduce reflections.

dBmV	Voltage	dBmV	Voltage	dBmV	Voltage
−60	1.00 μV	−20	100 μV	+20	10.0 mV
−59	1.12 μV	−19	112 μV	+21	11.2 mV
−58	1.26 μV	−18	126 μV	+22	12.6 mV
−57	1.41 μV	−17	141 μV	+23	14.1 mV
−56	1.59 μV	−16	159 μV	+24	15.9 mV
−55	1.78 μV	−15	178 μV	+25	17.8 mV
−54	2.00 μV	−14	200 μV	+26	20.0 mV
−53	2.24 μV	−13	224 μV	+27	22.4 mV
−52	2.51 μV	−12	251 μV	+28	25.1 mV
−51	2.82 μV	−11	282 μV	+29	28.2 mV
−50	3.16 μV	−10	316 μV	+30	31.6 mV
−49	3.55 μV	−9	355 μV	+31	35.5 mV
−48	3.98 μV	−8	398 μV	+32	39.8 mV
−47	4.47 μV	−7	447 μV	+33	44.7 mV
−46	5.01 μV	−6	501 μV	+34	50.1 mV
−45	5.62 μV	−5	562 μV	+35	56.2 mV
−44	6.31 μV	−4	631 μV	+36	63.1 mV
−43	7.08 μV	−3	708 μV	+37	70.8 mV
−42	7.94 μV	−2	794 μV	+38	79.4 mV
−41	8.91 μV	−1	891 μV	+39	89.1 mV
−40	10.0 μV	0	1.00 mV	+40	100 mV
−39	11.2 μV	+1	1.12 mV	+41	112 mV
−38	12.6 μV	+2	1.26 mV	+42	126 mV
−37	14.1 μV	+3	1.41 mV	+43	141 mV
−36	15.9 μV	+4	1.59 mV	+44	159 mV
−35	17.8 μV	+5	1.78 mV	+45	178 mV
−34	20.0 μV	+6	2.00 mV	+46	200 mV
−33	22.4 μV	+7	2.24 mV	+47	224 mV
−32	25.1 μV	+8	2.51 mV	+48	251 mV
−31	28.2 μV	+9	2.82 mV	+49	282 mV
−30	31.6 μV	+10	3.16 mV	+50	316 mV
−29	35.5 μV	+11	3.55 mV	+51	355 mV
−28	39.8 μV	+12	3.98 mV	+52	398 mV
−27	44.7 μV	+13	4.47 mV	+53	447 mV
−26	50.1 μV	+14	5.01 mV	+54	501 mV
−25	56.2 μV	+15	5.62 mV	+55	562 mV
−24	63.1 μV	+16	6.31 mV	+56	631 mV
−23	70.8 μV	+17	7.08 mV	+57	708 mV
−22	79.4 μV	+18	7.94 mV	+58	794 mV
−21	89.1 μV	+19	8.91 mV	+59	891 mV
				+60	1.00 V

μV = microvolts = 10^{-6} volts
mV = millivolts = 10^{-3}

Fig. 6.2 dBmV conversions to voltage levels.

CHAPTER 7

Broadband Passive Components

The most important component of any broadband communication system is the coaxial cable. As described in Chapter 4, the first CATV systems used open line feeder, not coaxial cable, and some FM aerial downleads still use open line systems. Coaxial cable is generally superior in low signal level applications, due to the much more effective screening against external electrical noise. A coaxial cable, as its name implies, consists of two concentric conductors, spaced by some insulation. The superior noise performance comes from the fact that the outer conductor or shield is always at ground potential with respect to any high frequency signals on the inner conductor. This means that external noise will be conducted by the shield to earth, and will have no influence on the inner conductor.

The drawback of coaxial cable, and the reason that open lines are still used in high power applications, is that the inner conductor must be spaced apart from the shield. Obviously, it must not be allowed to touch the shield. It can only be held away by using some form of insulating material, and this insulating material will have a higher dielectric constant than air – i.e. the capacitance of the cable will be increased, the dielectric loss will also increase, and the loss will increase, especially at high frequency. (The dielectric loss becomes important with very high power transmitters, as the dielectric heating can become a problem.) To cut down the problem of conductor spacing as much as possible, it is general practice to try to include as much air as possible in the cable, by using foamed polymers to support the inner conductor, or by using small discs every two or three centimetres.

As can be readily envisaged, the coaxial cable forms a type of capacitor. If the plate spacing of a capacitor is increased, then the capacitance will decrease. Exactly the same applies to coaxial cable – if the diameter of the coaxial is increased, then the capacitance between the inner and outer conductors will be reduced. The effect will be to reduce the attenuation at high frequencies. This is a general rule, and means that thin cables always attenuate more than thick cables.

Apart from attenuation, one of the other important parameters of a coaxial cable is the amount of screening against external electrical noise. In theory, for a solid pipe outer shield type of cable, where the shield is well grounded, this is a 100% effective screen. A large number of coaxial cables, however, use a screen made of wire braiding and/or metal tape simply wound around the dielectric. Such construction leaves physical 'holes', which act as electrical 'holes', through which noise can penetrate.

The noise immunity of a braided cable cannot be quoted for a range of cables, as it is dependent on the thickness of the braid weave and the amount of overlap if a metal tape shield is utilised. Hence a typical television aerial downlead, where noise immunity is not important, has a very loose weave, whilst high quality flexible coaxial can have up to four layers of braid and lapped metal tape.

Technically, we can define seven important characteristics of coaxial cables. They are:

(1) Impedance.
(2) Attenuation.
(3) Temperature characteristics.
(4) Loop resistance.
(5) Velocity of propagation.
(6) Noise immunity.
(7) Mechanical properties.

IMPEDANCE
Impedance is the effective AC resistance of a length of the cable. As the two conductors form a capacitor to earth, and an inductor along their length, they will behave like a resistor to any high frequency signal connected to a cable end. The value of this AC resistance is 75 Ω for CATV cables. This value is for an infinite length of cable, and short cables (a few metres) will not look like pure 75 Ω AC resistors. Any long length, however, will appear exactly like a pure resistor. This effect is important, as it allows broadband networks to function even in the presence of cable damage, provided the cable runs are reasonably long.

ATTENUATION
Attenuation is best quantified as shown in Fig. 7.1, but is basically the loss per unit length of an applied high frequency signal. It is usually measured in dB per metre, and the measurement frequency must be stated, as the effect is highly frequency dependent.

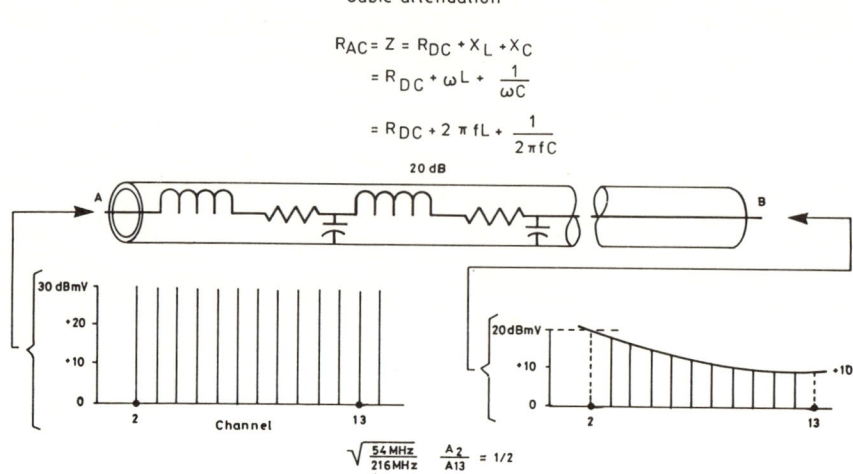

Fig. 7.1 Cable attenuation.

TEMPERATURE

Temperature characteristics are important on long cable runs, such as are found in CATV systems, but do not usually occur as problem areas in data networks. A coaxial cable's attenuation will change by about 1% for every 5°C change in temperature. Over a long cable run, extremes of temperature can cause quite severe changes in attenuation, easily enough to overload amplifiers in warm weather (attentuation decreases with temperature). Such problems are not found in short data systems in the UK, but can occur in the USA, on 20 km cable runs.

LOOP RESISTANCE

The loop resistance of the cable is important when it is wished to send power for amplifiers or other devices along the cable. It is simply the ohmic resistance of a length of the cable, and is invariably determined by the diameter of the centre conductor.

VELOCITY OF PROPAGATION

The velocity of propagation of a coaxial cable is the speed at which high frequencies travel on the cable. This is usually around 70–80% of the velocity of light in a vacuum. Velocity of propagation is critical if it changes with frequency, as then effects such as group delay, where one part of a signal arrives before another, become important. **Luckily, this effect is more pronounced on twisted pair cables, than on coax.**

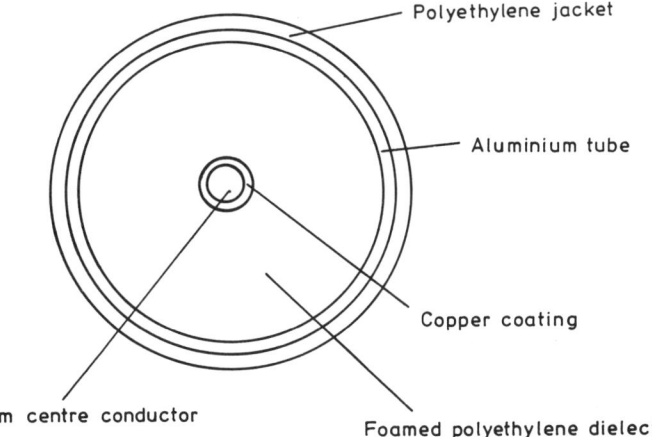

Basic cable data:

Impedance:	75 Ω
Attenuation:	4.3 dB per 100 at 300 MHz
Temperature characteristics:	Attenuation changes by 1% per 5°C
Loop resistance:	0.58 Ω per 100 m
Velocity of propagation:	88%
Noise immunity:	at 5 MHz, 110 dB
	at 300 MHz, greater than 1000 dB
Bending radius:	18 cm

Fig. 7.2 American standard 0.5 in. CATV coaxial.

NOISE IMMUNITY

The noise immunity has already been discussed, but mention should be made that immunity increases with frequency, for the cable shield looks more like a true ground as its dimension with respect to wavelength increases. Thus it is better to avoid using the whole of the short wave broadcasting frequency area (up to 30 MHz), as most coaxial cables are less noise immune in this area, and most noise sources (ignition systems, motors, thermostats, etc.) concentrate most of their energy in this range. In addition, the short wave has a large number of megawatt broadcast transmitters which contribute to the general noise level.

MECHANICAL PROPERTIES

The mechanical properties are the conventional properties of bending radius, weight, etc. It should be noted that no coaxial cable should be bent too sharply. Even if the centre conductor is undamaged, the

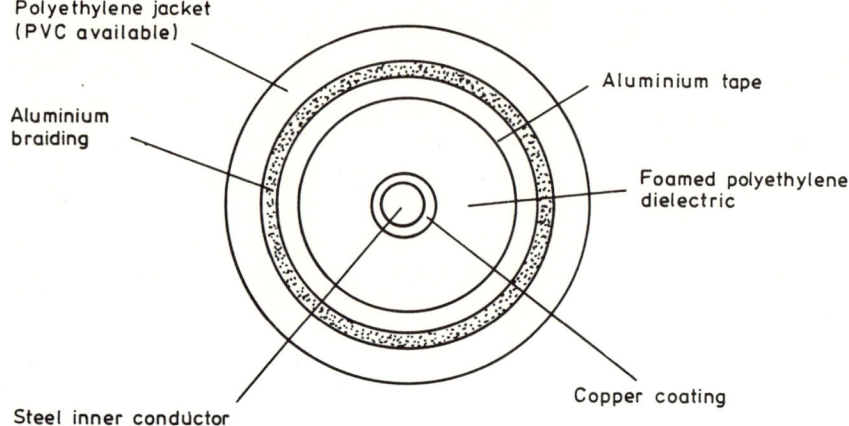

Fig. 7.3 RG 6 subscriber 'drop' cable.

Basic RG 6 data:

Diameter:	0.25 in
Impedance:	75 Ω
Attenuation:	11 dB per 100 m at 300 MHz
Temperature characteristic:	Attenuation changes by 1% for 5°C change
Loop resistance:	14.07 Ω per 100 m
Velocity of propagation:	81%
Noise immunity:	at 5 MHz, 60 dB
	at 300 MHz, 85 dB

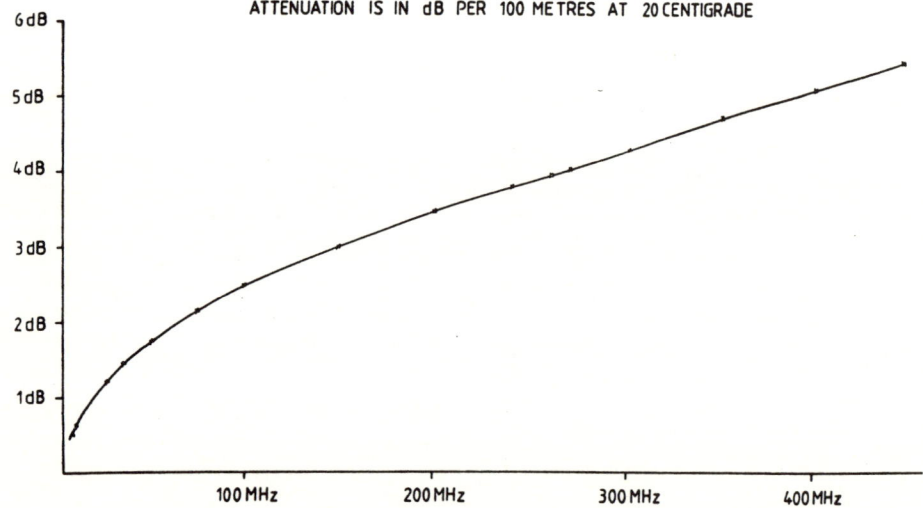

Fig. 7.4 Attenuation *vs.* frequency for 0.5 in. CATV cable.

outer shield will be partially ruptured, especially if that shield is of the bonded aluminium type, where a layer of aluminium foil is actually chemically bonded to the dielectric. Such damage will decrease the noise immunity. More importantly, any severe bend or kink on the cable will alter the capacitance value between the inner and outer conductors. This in turn will cause an alteration of the impedance at that point, causing a mismatch, with the attendant attenuation and reflections.

The type of cable used in a typical broadband data system can vary, but it is good practice to standardise on the best types of American CATV cable. We will examine the two main types of this cable in some detail. The seven points of performance are given in Figs 7.2 and 7.4 for 0.5 in. trunk cable and in Figs 7.3 and 7.5 for RG-6 subscriber drop cable.

CABLE CONNECTORS

There are two connector systems in general use on CATV systems. On the subscriber side, the F connector is used, and a robust, watertight coaxial connector on the trunk cables.

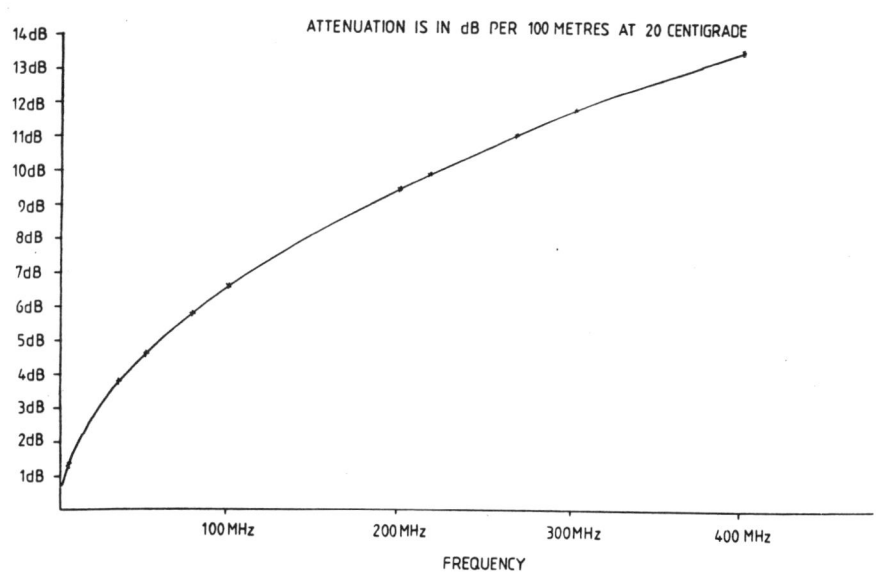

Fig. 7.5 Attenuation *vs.* frequency for RG 6 'drop' cable.

The F connector appears rather flimsy to people used to BNC connectors, as it uses the inner conductor of the cable as the connector centre pin. This is possible because the centre conductor of the RG-6 drop lead is made of steel. The steel is more or less inextensible and remains the same diameter after the cable has been pulled or stretched. It will thus provide a 'spring' force inside the mating half of the connector, providing a gastight connection for many years. This is a more important point than is often realised – if the centre pin were copper, it would relax under the spring force of the wiping leaf in the mating part of the connector and with time would allow air to oxidise the connecting surfaces, leading to rather odd RF problems (partial rectification, etc.). Equally important is that the connector is matched to the cable inner conductor diameter – this should not be so large as to strain the wiping leaf apart. If this should happen, the first connection will be fine, but if this connection is changed, then the next user might find that the socket has gone intermittent.

A great advantage of the F cable connector (male) is the price, literally pence each. The speed of fitting, with the appropriate crimping tool, is also impressive. If large numbers of drop cables must be terminated, there are electrically driven machines available which can prepare the coax, feed the connector and crimp.

The 0.5 in. trunk cables are not designed to have mid-span connectors, but should run from tap to tap to amplifier in continuous runs. The taps and amplifiers have a standard connector fitting, consisting of a ⅝ in. threaded hole, with a sort of 'chocolate block' type connector in the centre, set some way back into the body of the device. The simplest form of connector is a tapered, threaded cap, through which the outer conductor passes (see Fig. 7.6). Screwing the cap into the device forms a connection to the outer sheath. The centre conductor can be gripped by the 'chocolate block'.

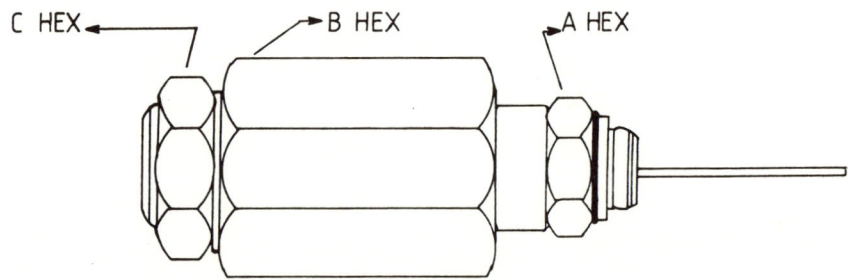

Fig. 7.6 Connector.

As described, this is a very poor system (though still used), as there is no spring force in the connection, and corrosion will soon be able to set in. The best systems today use a separate connector to fit into the trunk devices. These connectors have integral steel mandrels onto which the outer shield is compressed, whilst the inner conductor is forced by the tightening of the connector into a steel cone. On the end of this cone is a long steel pin, which will be the connection to the 'chocolate block' screw terminal in the trunk device.

All the screw threads on such a connector or 'sting' have 'O' ring rubber seals. It should be noted that there are very great variations in the quality of American CATV connectors. These form the weak links of any broadband cable system, and the best should be installed. The extra cost is negligible, and the trouble that, for instance, can be caused by a relaxed shield connection starting to corrode can be very difficult to cure in the field.

Due to the age of the CATV industry, there are a host of connector variations on the market. Apart from the straight 'sting' product, there are in-line connectors, and 90° corner joints, both helpful in building installation. A wide range of adaptors between the ⅝ in. thread system and the F connector (see Fig. 7.7) are available. Figure 7.8 shows an F connector terminating resistor used to terminate unused cable outlets.

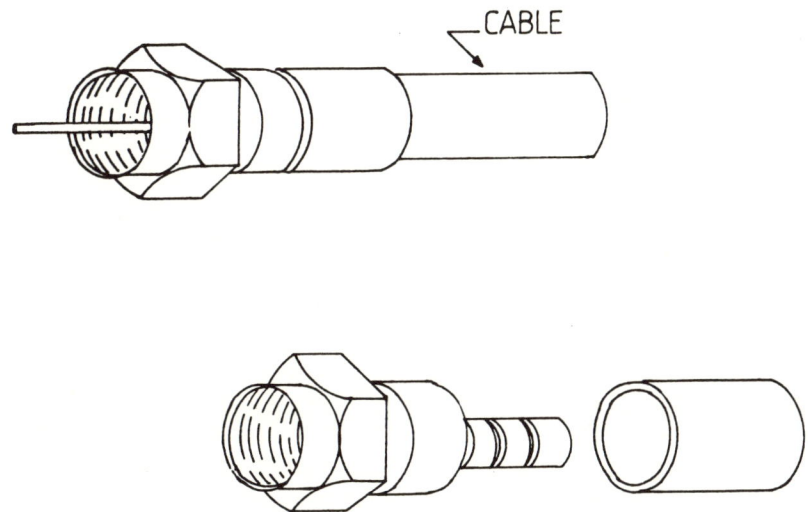

Fig. 7.7 'F' connector.

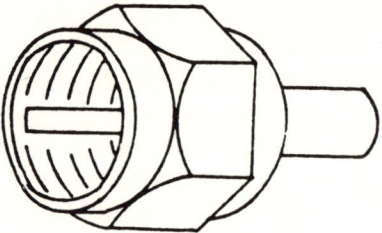

Fig. 7.8 'F' terminating resistor.

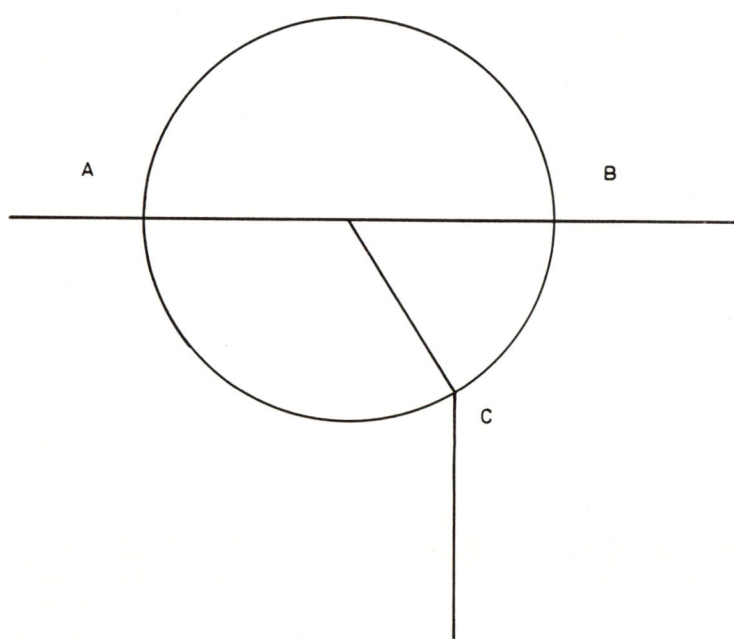

A - B = Insertion loss
A - C = Attenuation value
B - C = Isolation
For serious system use, isolation should be greater than 30 dB and return loss should be greater than 20 dB.

Fig. 7.9 The directional tap or splitter.

SPLITTERS AND TAPS

Splitters and taps can be divided into four separate groups:

Trunk cable splitters.
Trunk cable taps.
Miniature splitters.
Miniature taps.

As has been described, all taps in CATV use today are transformer coupled. The words 'splitter' and 'tap' are interchangeable, as the electrical functions are identical. The only difference is the output connectors. All these devices operate by splitting the available signal power between a main cable and a drop cable (for a tap) or another spur of the main cable (for a splitter), the diagrammatic representation of which is shown in Fig. 7.9.

The important parameters of taps and splitters are:

Insertion loss.
Attenuation value.
Isolation.
Return loss.

The insertion loss is the loss in dB across the device in the direction of the main cable – how much of the signal is lost on the main trunk.

The attenuation value is the loss in dB to the subscriber drop or spur cable. The insertion loss and the attenuation are interrelated – the lower the attenuation value, the higher the attenuation in the trunk cable, i.e. the insertion loss. This is because the power is literally split between the through direction and the drop or spur output.

The isolation is the attenuation in dB over the blocked reverse path to the drop or spur. As high a figure as possible should be sought here, in any event greater than 30 dB.

The return loss is a measurement of the quality of the match of the device to the cable system, and is the ratio of the amount of power reflected by the device. A return loss of 20 dB or greater should be looked for.

It can be difficult to describe splitters and taps in detail, due to variations between manufacturers. As a guide to the different types of splitters and taps we can consider the Magnavox range.

Trunk cable splitters are available in five attenuation values – 3.9, 8.0, 12.0, 16.0 and two times 7.0 dB. Input and output(s) are the ⅝ in. CATV threaded connection. The use of such splitters is to form spurs and branches on the trunk cable network.

Trunk cable taps have ⅝ in. CATV input and output connectors for the main cable, but have F connector drop outputs. They are

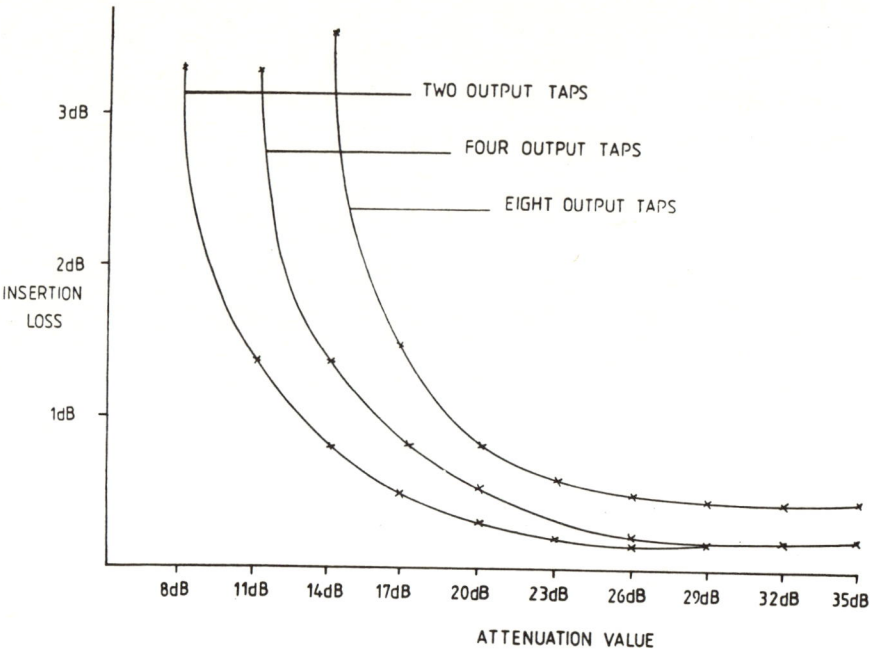

Fig. 7.10 Insertion loss *vs.* attenuation value for trunk cable taps.

available in two, four and eight drop versions. Attenuation values from 8 dB to 32 dB in 3 dB steps are provided. The insertion loss depends on the attenuation value and the number of F drop ports. A graph of tap attenuation value against insertion loss is given in Fig. 7.10.

Miniature splitters are available for splitting signals on drop cables, and have F connector inputs and outputs. Splitters with attenuations of 3.5, 7.0 and 10.0 dB are available, being respectively two-, four- and eight-way splitters. These devices are very small – about the size of a matchbox.

Miniature taps are the same as the miniature splitters, except that they are taps. Attenuation values in four dB steps from 6 dB are available. They are usually single-tap output only.

It can be seen that the range of taps and splitters is very wide indeed, as other suppliers have extensions to this basic range in some areas. It should be noted, however, that the isolation and return loss of some of the miniature taps and splitters leaves something to be desired. They would only be used in a CATV system at the end of a drop cable, in the subscriber's premises, to extend the system to other television sets in the house. It is probably unwise to base the design

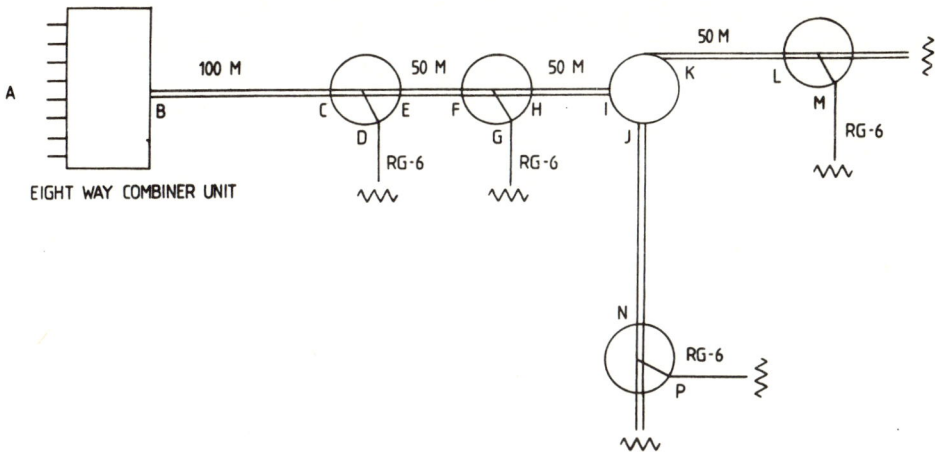

Device data:
  Two output taps:
    attenuation: 8 11 14 17 20 23 26 29 32
    insertion: 3.3 1.4 0.8 0.6 0.3 0.3 0.2 0.2 0.2
  Splitters:
    attenuation: 3.9 8 12 16
    insertion: 3.9 2 1.2 1
  Eight-way combiner: attenuation all ports, 10 dB
  0.5 in. CATV cable: 4.3 dB per 100 m
  RG 6 CATV drop cable: 11.7 dB per 100 m

Fig. 7.11 Passive system design example.

of an entire broadband system on these devices.

A variation on the miniature taps for distribution in a small area are the television type outlet plate taps. These are used extensively in blocks of flats, etc., but cannot be recommended for serious broadband network use. They are usually of the resistive tap type, and have one or at most two attenuation values. Whereas television receivers are insensitive to signal levels, data systems are not (mainly due to the transmitter levels on the reverse path), and it is not possible to do serious system design with these components.

PASSIVE SYSTEM DESIGN

The design aim for the passive system shown in Fig. 7.11 is to achieve a constant output level at the end of each drop lead. For the purposes of this design, this level should be 0 dBmV. A transmitted level of +45 dBmV can be assumed for the system headend.

Location	Level
A	+45
B	_____
C	_____
D	_____

Tap chosen = _____
Level at end of drop = _____

D	_____
E	_____
F	

Tap chosen = _____
Level at end of drop = _____

G	_____
I	
M	_____
N	

Splitter chosen = _____

Tap chosen = _____
Level at end of drop = _____

J	_____
K	_____
L	

Tap chosen = _____
Level at end of drop = _____

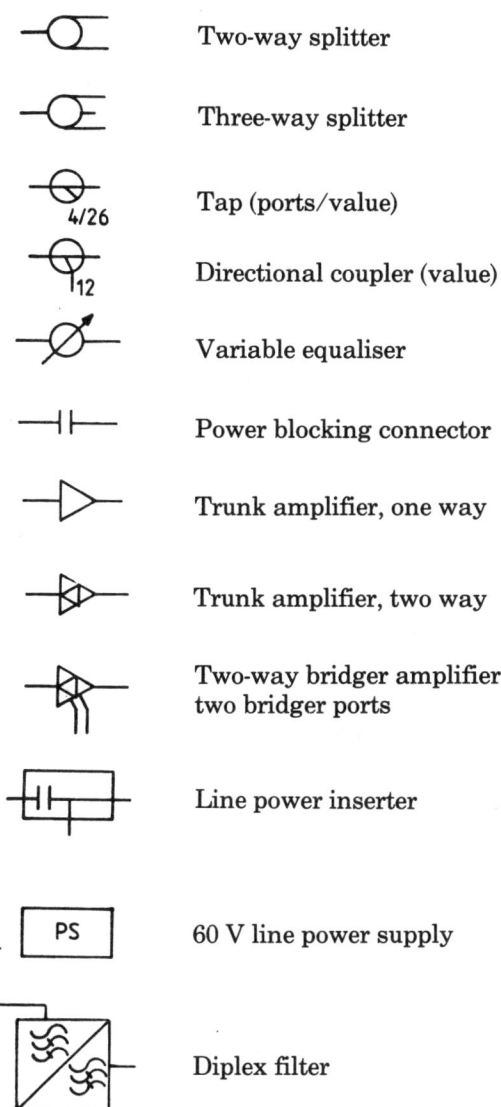

Two-way splitter

Three-way splitter

Tap (ports/value)

Directional coupler (value)

Variable equaliser

Power blocking connector

Trunk amplifier, one way

Trunk amplifier, two way

Two-way bridger amplifier
two bridger ports

Line power inserter

60 V line power supply

Diplex filter

Fig. 7.12 Symbols used for cable system components.

CHAPTER 8

Broadband CATV Amplifiers

The broadband amplifier forms the heart of a CATV system – without amplification, cables are restricted to a few hundred metres. It is not generally realised just how many amplifiers are used in a typical system – up to 2000 is not uncommon. This is because the design of a city-wide CATV system is rather different from a typical data system.

There are four main types of amplifier: *trunk, bridger, line extender* and *distribution amplifier*. These are used in different parts of the CATV network, and have differing electrical specifications.

Trunk amplifiers

Trunk amplifiers are the very high quality main amplifiers on the 'backbone' trunk cable system. This cable backbone will be laid around the town in the normal 'tree' fashion. There will be no taps in this backbone, just cable, splitters and amplifiers. As far as possible, the cable will run from trunk amplifier to trunk amplifier, the amplifiers providing exactly enough gain to balance the attenuation of the cable run. (This is called the 'unity gain concept', whereby the overall system gain never exceeds unity.)

Bridger amplifiers

When service is needed at any one point, a bridger module is inserted in the trunk amplifier – this is, in effect, another totally separate amplifier, on the end of a splitter. The output of the bridging module is led to a cable run which will have subscriber taps on it. In this way, the main cable run is protected from the cable runs passing near the subscribers – any cable damage on the subscriber run will only affect that run, not the backbone system. The cable near the subscribers is much more likely to be damaged than the backbone, with a large percentage of trouble caused by illegal connections and malicious experimentation. This problem is not understated – the unscrambled

CATV service in residential areas of Dublin, Eire, has on average 10% illegal connections at any one time, on a system with 20 000 subscribers. These connections are often made by filing away the outer shield of the cable in order to tap onto the centre conductor.

Line extender amplifiers

The length of cable from a bridging amplifier is restricted by the number of taps on that cable, as each tap on the cable will bring with it insertion loss. If the signal level falls below the necessary level, then the line extender amplifier is used, to boost the level. It is an inexpensive, lower quality unit, which is not designed to run in cascades, but functions exactly as its name implies – it extends the subscriber cable line.

Distribution amplifiers

The last type of amplifier is the distribution amplifier. This is also an inexpensive unit, designed to boost the signal for distribution in blocks of flats, etc.

Figure 8.1 shows the use of the different types of amplifier in a CATV system, conventionally a 'forward only' system. Amplifiers in a data or mixed system must be bidirectional, and this is implemented by using two amplifiers in each position, one for each direction. The frequency spectrum is split according to whether the system is sub- or mid-split, by diplex filters in the station housing. Because of the need to add extra filters and another amplifier module for the reverse channels, the only units that offer mid-split reverse channels today are trunk amplifiers. (It should be mentioned that the units are highly modular, and it is possible to plug all sorts of bridger, reverse and filter modules into a main trunk amplifier housing.)

At the time of writing, no manufacturer offers line extenders for mid-split operation, although these would be useful in many applications. Some sub-split line extender equipment exists.

The chief function of an amplifier is to provide gain on a system, to overcome the cable and insertion losses in that system. It is an accepted principle that an amplifier merely restores the losses, and does not add to the gain of the system – the unity gain concept. This leads to the establishment of a fixed loss in a design, after which an amplifier is used. By convention, this loss is between 20 and 26 dB, and is called the *spacing* of the system. Thus a system designer will talk about amplifiers 'spaced' every 22 dB. This means that the signal level on the cable trunk will have decreased by 22 dB at the input of the next amplifier. If there were only trunk cable between the

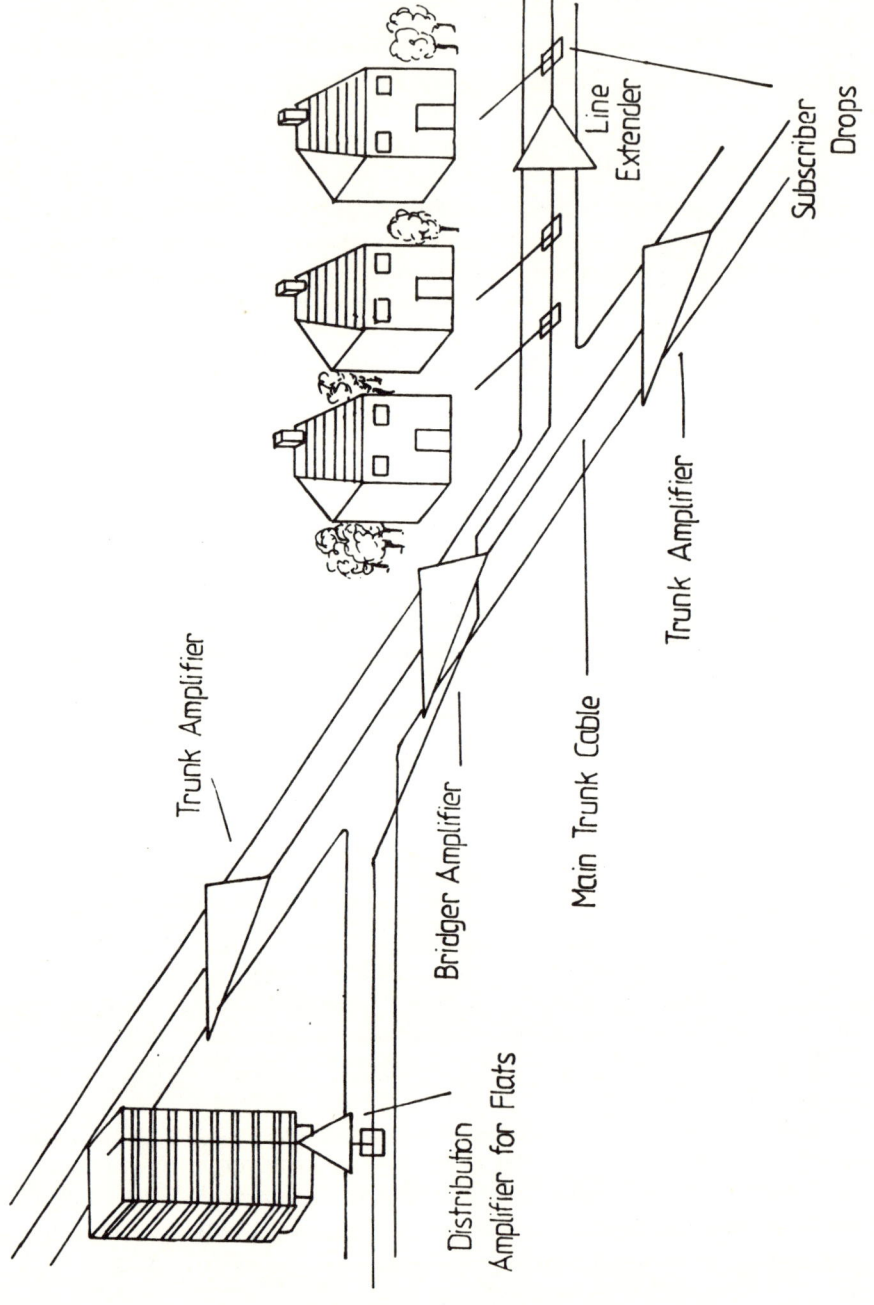

Fig. 8.1 Amplifiers in a CATV system.

amplifiers, a 22 dB spacing would be 512 metres (4.3 dB per 100 m). The trunk amplifiers that are available comply with this *ad hoc* standard, and commonly have maximum gains of 27–28 dB. This means that a design spacing of 22 dB is conservative, in that it allows for the later insertion of a splitter on the trunk between amplifiers.

Because of the standard design practice, the gain control range of an amplifier is quite limited – around 8 to 10 dB for many amplifiers. This means that if only 15 dB gain is required from an amplifier, it will not be possible to set this up using the built-in gain controls. To solve this sort of problem, most station housings have slots for the insertion of external attenuator pads. A typical use of such pads would be to reduce signal level to the input of an amplifier on a system that will be extended at some future time by insertion of splitters or taps in the trunk cable run.

Noise

It is unfortunately a fact that real world amplifiers do not just add gain to a system – they introduce noise and distortion. These become

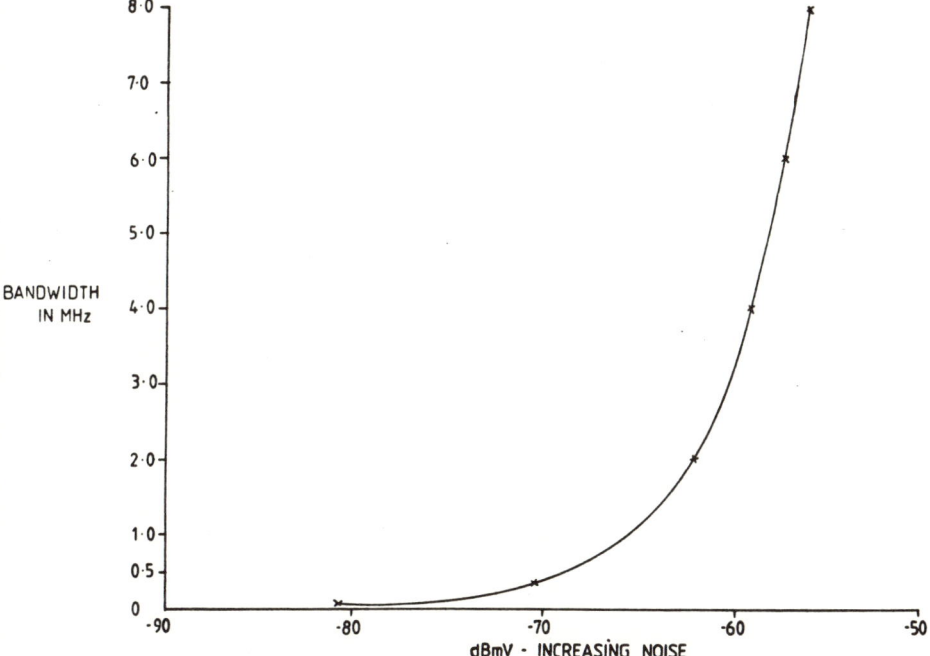

Fig. 8.2 Thermal noise *vs.* bandwidth.

large factors when designing big systems with cascades of amplifiers, and in fact become the most important system design parameters.

Noise, in the electrical sense, is thermally generated. As the cable system looks like a 75 Ω resistor, it will generate an amount of thermal noise. The actual amount of noise that we measure depends on the bandwidth of the noise receiver. Noise is distributed across the entire frequency spectrum of the cable, and the wider the receiver bandwidth, the more noise will be picked up (see Fig. 8.2). As the noise is thermal, there will be more noise at high temperatures, but for practical purposes this variation can be ignored. In absolute terms the thermal noise on a piece of cable over a 6 MHz television channel will be –57.4 dBmV, which corresponds to a voltage of 2.6 μV. Table 8.1 shows bandwidth *vs.* noise generated.

Any practical amplifier will generate its own noise, due to the various semiconductor noise sources inside the circuitry. The amount of noise it generates over the cable thermal noise is the noise performance of the unit. This is expressed as a ratio, in dB, and is known as the *noise figure* of the amplifier. As low a noise figure as possible is desirable. The amplifier will always amplify the noise on the cable as well, so when considering the output of an amplifier, this must be taken into account.

To work through an example:

Noise on cable (6 MHz channel):   –57.4 dBmV
Noise figure of amplifier: 10 dB (practical value)
Gain of amplifier: 23 dB
Noise output of amplifier = noise + gain + noise figure
                                     = –57.4 + 23 + 10
                                     = –24.4 dBmV

By itself, the noise output of the amplifier is not too important, but the ratio of the signal to that noise voltage is. Here we are stepping into a fairly complicated area, so we will consider only carrier-to-noise ratio, rather than signal-to-noise. For most data systems, the carrier-to-noise ratio is the more important, as the signals themselves are hard to analyse. Data devices on cable systems invariably

Table 8.1 Thermal noise *vs.* bandwidth

Bandwidth (MHz)	4.2	.03	.3	2	6	88	4
Impedance (Ω)	75	75	75	75	75	75	75
K (Boltzman's Constant)	4–15	4–15	4–15	4–15	4–15	4–15	4–15
En (Noise voltage)	2.245–6	1.897–7	6.000–7	1.549–6	2.683–6	3.098–6	2.191–6
X	1.122497	0.948683	0.3000000	0.7745967	1.341641	1.549193	1.095445
dBmV	–58.9963	–80.4576	–70.4576	–62.2185	–57.4473	–56.1979	–59.2082

radiate a pure carrier when not transmitting data, and it is this carrier level that we shall consider in relation to carrier-to-noise studies.

As has been shown, carrier-to-noise ratio is an expression of the ratio of the carrier voltage level to the noise voltage level in dB. It is a measurement of the 'freedom' of carrier level within that system.

Taking our last example:

Amplifier gain:	23 dB
Noise figure:	10 dB
Thermal noise (6 MHz):	–57.4 dBmV
Carrier signal in:	17 dBmV

Hence
Carrier output signal = 40 dBmV

From last example,
Noise output = –24.4 dBmV

Thus,
Carrier-to-noise ratio or C/N = 64.4 dB

Distortion in amplifiers is rather harder to quantify for data devices applications, as it is impossible to calculate the amount of distortion from the published specifications, which relate to television signals. As a general basis, it is recommended that the calculations for distortion are done using the maximum possible number of TV channels on the bandwidth. If these distortion calculations provide an adequate freedom from distortion products for the television channels, it may be assumed that the system will be adequate for data use. (This may seem empirical, but with data networks such as Sytek LocalNet, the number of carriers changes statistically, and the true relationships of the distortion products would be unpredictable.)

Distortion is the result of non-linear amplification, whereby some parts of the incoming signal will be amplified more than others. Thus an incoming sinewave signal will have a different shape at the output. This signal of different shape is composed, mathematically, of a whole series of sinewaves at different frequencies, with, however, the main sinewave still at the original frequency. The other sinewaves are known as the 'harmonics' of the original wave, and will be found at various multiples of the original frequency. The process is further complicated when more than one signal is amplified at the same time, when sum and difference frequencies will also be generated. With a large number of carriers on a system, the number of such distortion products can become so high that they look like a continuous noise spectrum on the system.

Another form of distortion, related to the harmonic and beat distortion, is cross-modulation, which is when one television signal will get partially modulated by another. This effect is impossible to relate directly to data systems, but forms an important measurement of the third order distortion products. Distortion is classified as second order, which includes harmonic distortion, and third order, which includes cross-modulation. In practical amplifiers, second order distortion is very low, but third order distortion can be a problem. (See figs 8.3, 8.4 and 8.5.)

Examples

The following examples use some formulae included without proof, in order to facilitate the reading of amplifier specifications.

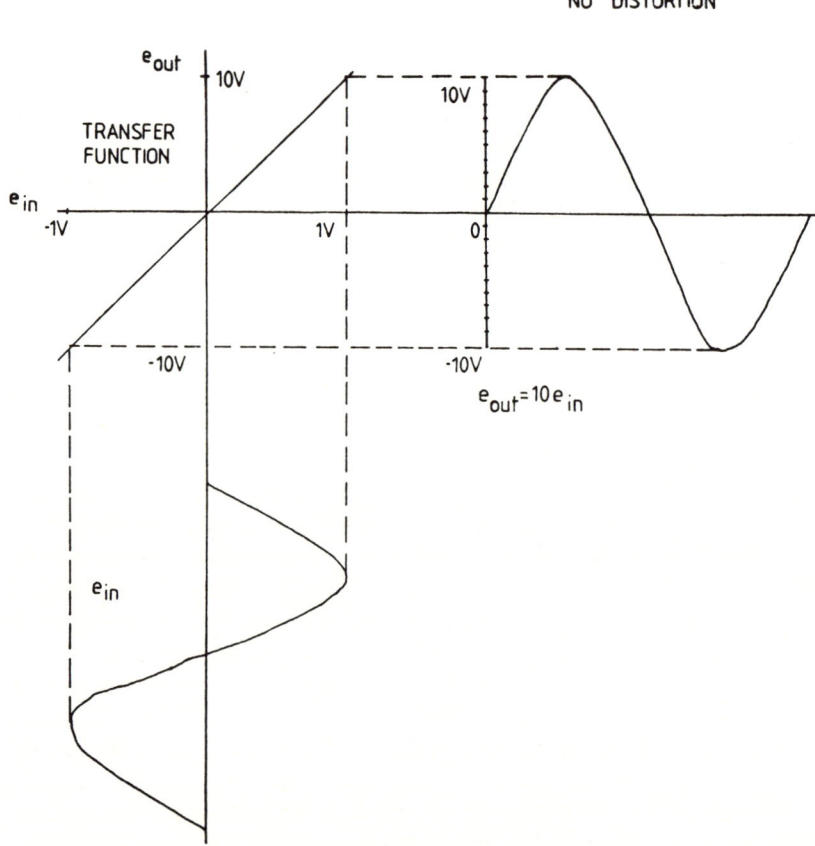

Fig. 8.3 Amplifier distortion.

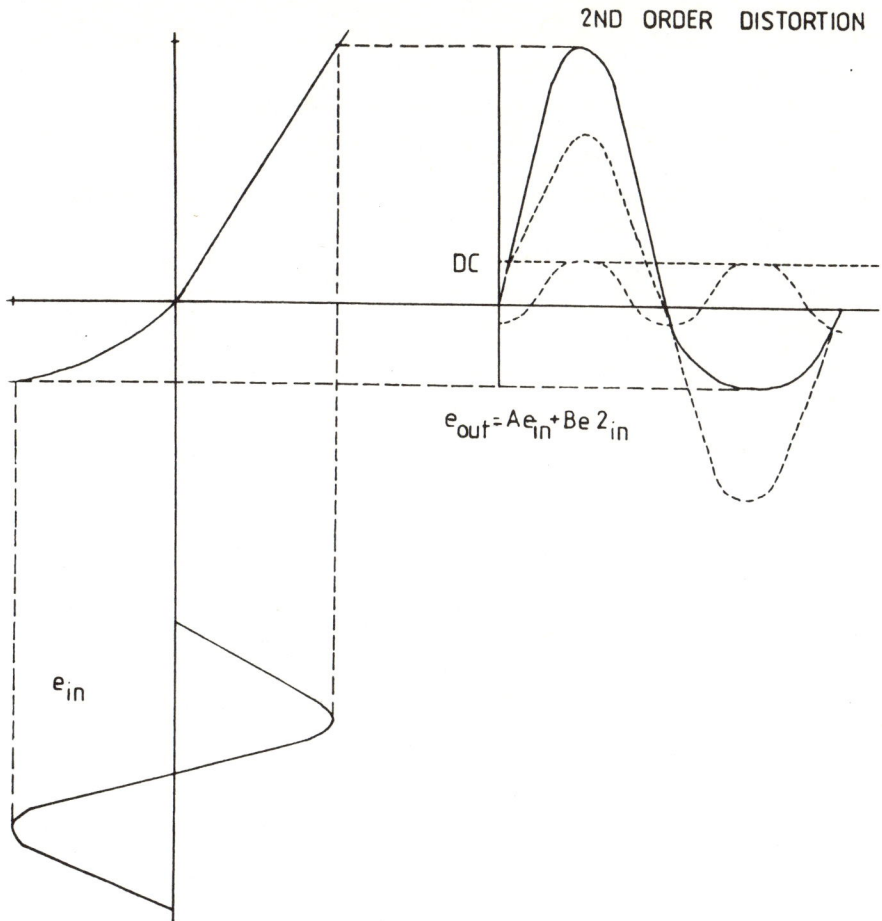

Fig. 8.4 Amplifier distortion, 2nd order.

Three specifications should be found when considering distortion products:

Cross-modulation level.
Second order beats.
Importantly, the output level for the above two figures.

Second order products are so called because they are mathematically related to the square of the input voltage. This means that raising the input voltage to the amplifier by 1 dB will raise the second order products by 2 dB. Similarly, third order products, including cross-modulation, are related to the cube of the input voltage, and raising this by 1 dB raises the third order products by 3 dB. In addition, the cross-modulation level is proportional to the number of channels on the cable.

A few examples, first of second order distortion.

Amplifier specification: –85 dB at 30 dBmV output level

Hence, the actual second order distortion is at (30–85) = –55 dBmV.

If we raise the output level to +45 dBmV, then the increase in system level is 15 dB or a 30 dB increase in the second order products, i.e. level = –25 dBmV.

For third order products and cross-modulation, the products

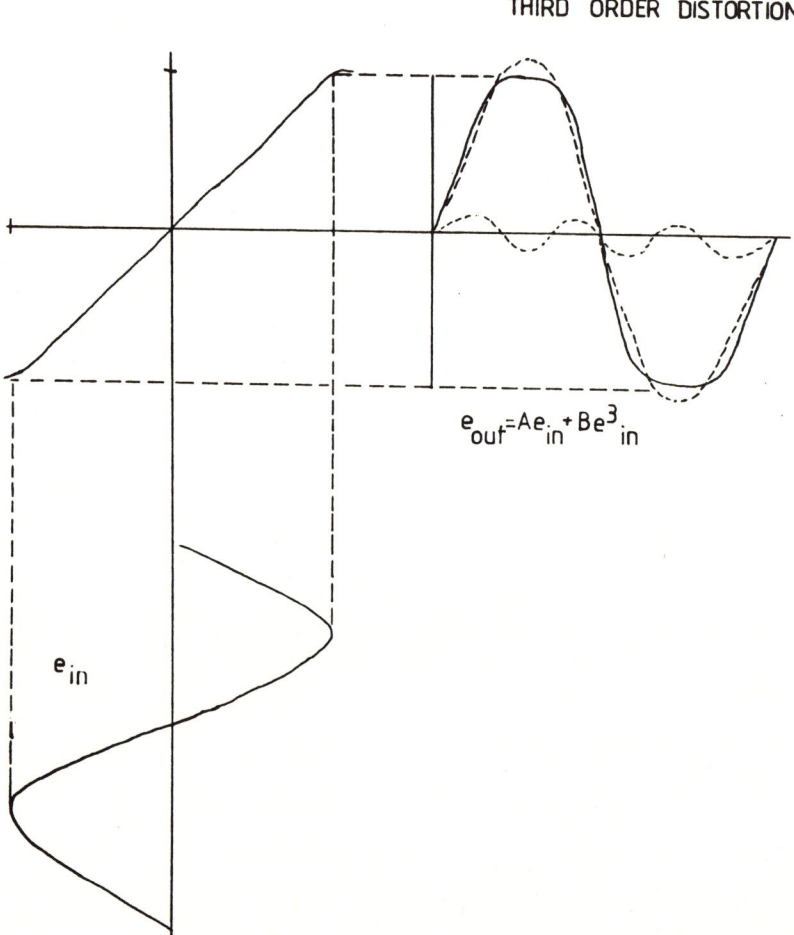

THIRD ORDER DISTORTION

$$e_{out} = Ae_{in} + Be^3_{in}$$

e_{in}

Fig. 8.5 Amplifier distortion, 3rd order.

increase by 3 dB for a 1 dB change in output level. Thus, for a cross-modulation specification of -91 dB at 30 dBmV output:

Cross-modulation level $= -61$ dBmV

If we raise output level by 15 dBmV to 45 dBmV, then the cross-modulation will increase by 45 dB, i.e.

Cross-modulation at 45 dBmV $= 16$ dBmV

Clearly, the third order products are much more important than the second order.

So far we have dealt with both second and third order products as absolute levels. They are normally quoted as ratios, and, as the system level in both our examples was raised by 1 dB, then for second order products it can be seen that the ratio gets 1 dB worse, and the cross-modulation ratio gets 2 dB worse.

Returning to the first example:

at 30 dBmV output, second order ratio was -85
at 45 dBmV output, second order ratio was -70

For each dB increase in system level, second order products ratio gets one dB worse.

For the second example:

at 30 dBmV output, cross-modulation ratio was -91
at 45 dBmV output, cross-modulation ratio was -61

For each dB increase in system level, cross-modulation products ratio gets 2 dB worse.

The effect caused by television channel loading on the cross-modulation ratio is given by the formula:

$$20 \log (N2 - 1/N1 - 1)$$

where $N2$ is the number of channels required, and $N1$ the number of channels that the cross-modulation is known for.

Problem

The following problem uses actual data from Texscan Phoenician II low cost trunk amplifier series (available as forward only or sub-split).

Amplifier gain set to:	25 dB
Input level at:	+17 dBmV
Noise figure:	7 dB
Second order beats:	-85 dB
Cross-modulation:	-95 dB

Cross-modulation and second order distortion measured at output level of:	+30 dBmV

Calculate:

(1) Carrier-to-noise ratio, C/N.
(2) Second order distortion level.
(3) Cross-modulation level.
(4) Second order distortion ratio.
(5) Cross-modulation ratio.

Thermal noise on the cable should be taken as –57.4 dBmV over 6 MHz bandwidth.

SOLUTION

(1) Noise output of amplifier = noise + gain + noise figure
$$= -57.4 + 25 + 7$$
$$= -25.4 \text{ dBmV}$$
Carrier output level = input level + gain
$$= 17 + 25 = 42 \text{ dBmV}$$
Hence C/N = 42 + –25.4 = 67.4 dB

(2) *Second order products*: actual level at the specified output of 30 dBmV = level – ratio = 30–85 =–55 dBmV.
To change this to allow for the higher output level, we increase this level by 2 dB for every 1 dB increase in the carrier output level.
Carrier output is 12 dB above the specified level, therefore the second order products are 24 dB higher.
Hence, second order product level =–55 + 24 =–31 dBmV.

(3) *Cross-modulation level*: actual level at the specified output of 30 dBmV = level – ratio = 30–95 =–65 dBmV.
To change this to allow for the higher output level, we increase this level by 3 dB for every 1 dB increase in the carrier output level.
Carrier output is 12 dB above the specified level, therefore the cross-modulation is 36 dB higher.
Hence, cross-modulation level = –65 + 36 = –29 dBmV.

(4) *Second order product ratio*: this is the ratio of the absolute second order product level to the carrier output level.
Ratio = carrier output – second order level = 42–(–31) = 73 dB.

(5) *Cross-modulation ratio*: the calculation is identical to (4).
Ratio = carrier output – cross-modulation level = 42 –(–29) = 71 dB.

EQUALISATION

We have seen that the cable used in CATV systems does not have constant attenuation with frequency, and that, although the slope of the attenuation characteristic is not steep at high frequencies, it can become very sharp at low frequencies. This is one of the reasons that the first forward-only systems started at 47 MHz and not at lower frequencies – the attenuation of the cable is so low at 5 MHz that there would be a problem at the input to amplifiers, and eventually subscribers, with the level being too high. Equalisation is the means of correcting this problem, by providing frequency sensitive passive circuits which attenuate more at low frequencies than at high, thus balancing out the cable attenuation. Very often such circuits are provided as plug-in modules on amplifiers.

Although the equalisation problem must be regarded on CATV systems, it is much more critical on data systems. A typical data system uses the low frequency end of the cable for the reverse channels, i.e. back towards the headend. In general, the system is designed to provide a constant output level at each drop, but this is designed at some frequency in the forward frequency range, i.e. a high frequency, with more cable attenuation. On the reverse path, the cable will have much less attenuation, and the signal arriving at the headend will be too high. It will be no good turning the transmitter level of the data device down, as it might have to be used close to the headend, where the cable content of the signal path is low, and the received signal at the headend would then be insufficient.

The best way around this problem is to insert passive cable equalisers at regular intervals throughout the cable run. It should be noted that equalisation problems will be more pronounced on systems using thin, high attenuation cable than on systems using 0.5 in. CATV trunk cable.

The equalisers in most amplifiers are adjustable, and can take two forms: the *passive equaliser*, which is just a frequency sensitive pad, and the *slope controlled amplifier*, in which the amplifier itself is made frequency sensitive and will have a 'slope' control, allowing extra gain to be added at high frequencies. A slope controlled amplifier is probably easier to use, as it does not introduce any overall loss on the system, and usually has a wider gain control range.

AMPLIFIER CASCADES

Any large practical system will have more than one amplifier in a cascade, and a set of rules exists as to how the noise and distortion effects accumulate in the cascade. The process is not strictly additive, as some of the effects are random in nature, such as thermal noise, and can cancel out in part in the next device.

It is the cascaded noise and distortion in a CATV system that determines the ultimate length of that system, and the amplifier quality that can be used. The reason why the inexpensive line extender amplifier cannot be used in cascades is that its noise and cross-modulation, combined with the fact that it operates at high output levels and high gain (sometimes up to 40 dB), mean that the output levels of distortion and noise cannot be further increased without causing signal degradation.

For data systems, in the absence of other specifications, the following television standards apply for a complete system:

Overall system C/N
should be greater than: 43 dB in a 6 MHz bandwidth
Overall system
cross-modulation should be: better than 52 dB

Without proof, the following cascade factors apply for amplifiers:

Carrier-to-noise:

C/N cascade = C/N amplifier – 10 log n

where n = number of amplifiers in the cascade. (Note: this assumes that the amplifiers will all have the same C/N, i.e. all be working at the same levels and gain. In practice, this is usual for a large system, as it facilitates maintenance.)

Second order products:

= –D amplifier + 10 log n (theory)
= –D amplifier + 15 log n (practice)

where D is the second order distortion ratio for the amplifier, and n is the number of amplifiers. Again, this assumes that all amplifiers in the cascade are similar. (Note: log is log to base 10 throughout.)

Third order products and cross-modulation:

= –D amplifier + 20 log n

again, where D is the cross-modulation ratio for one amplifier, and n is the number of units in the cascade.

Using these formulae, we can calculate the C/N and cross-modulation of any cascade of similar amplifiers. Using the results of the Texscan Phoenician amplifier for a cascade of 10 amplifiers, we get the results:

$$\text{C/N of single unit} = 67.4 \text{ dB}$$

C/N of cascade = C/N single unit $-$ 10 log (10)
$$= 67.4 - 10$$
$$= 57.4 \text{ dB}$$

Thus C/N is well within our specification of 43 dB.

$$\text{Cross-modulation of single unit} = 71 \text{ dB}$$

Cross-modulation $=-D$ single unit $+$ 20 log (10)
$$=-71 + 20$$
$$=-51 \text{ i.e. } 51 \text{ dB ratio}$$

This is just under the system specification of 52 dB, and must be regarded as marginal.

This result demonstrates the system design possibilities, as we

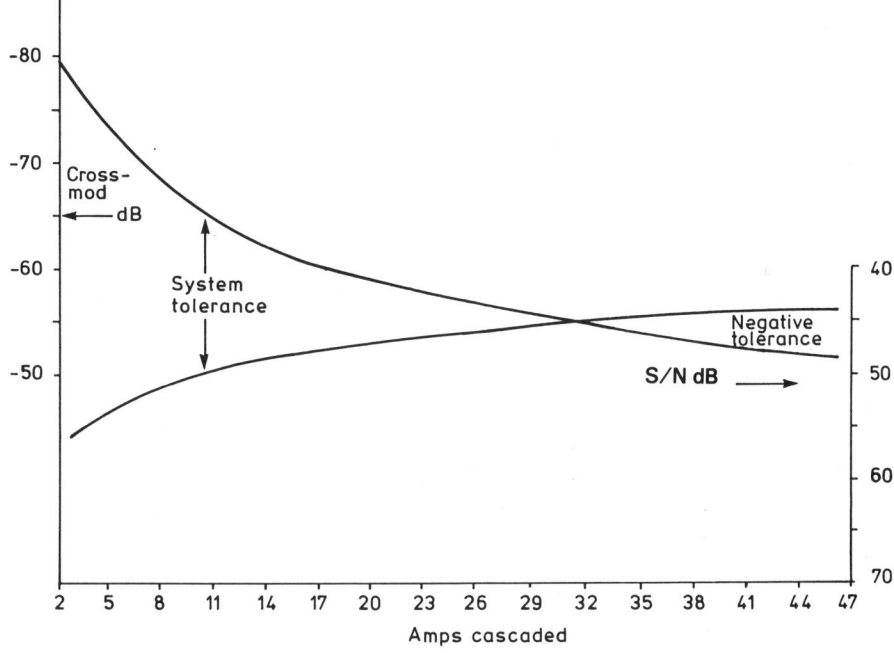

Designed for zero tolerance at cascade = 32
Given: NF = 10 dB
Trunk specification S/N = 45 dB
                    O/C = + 46 dBmV at $-$ 57 dB cross-modulation
Trunk cross-modulation specification = $-$ 55 dB
Amplifier input + 11.05 dBmV; amplifier output = 31.95 dBmV

Fig. 8.6 Trunk system tolerance as a function of amps cascaded.

have a marginal cross-modulation figure, but a good noise level on the cascade. It is therefore possible to decrease the system levels throughout the cascade, by 2 dB or so. This would decrease the carrier-to-noise ratio, but would improve the cross-modulation. There is thus a system tolerance, or tradeoff between cross-modulation and noise – if we decrease the system levels, the C/N will get worse; if we increase them, the cross-modulation will get worse. The system tolerance can be measured as the amount of 'freedom' we have to reduce levels before noise problems begin (see Fig. 8.6). In the example above, the freedom is $57.4 - 43 = 14.4$ dB, i.e. the difference between the C/N of the cascade at the current operating level and the recommended 43 dB C/N worse case specification.

It should be noted that for a typical local data system, such a long cascade would not be typical, but such cascades occur in CATV systems.

Cascades of different amplifiers

We have considered the case of similar amplifiers in cascade, but there are instances where this is not adequate for correct evaluation of system performance. The cases are where line extenders are added to trunk or bridging amplifiers, and, more interesting for data systems, where a remodulator is used at the headend to provide a 'broadcast' type of local area network. In the latter case there are two

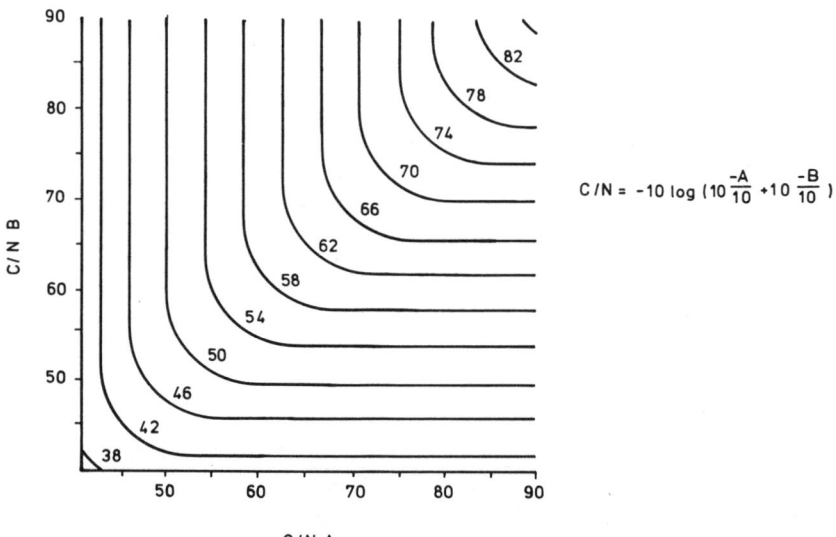

$$C/N = -10 \log \left(10^{\frac{-A}{10}} + 10^{\frac{-B}{10}}\right)$$

Fig. 8.7 Combining C/N.

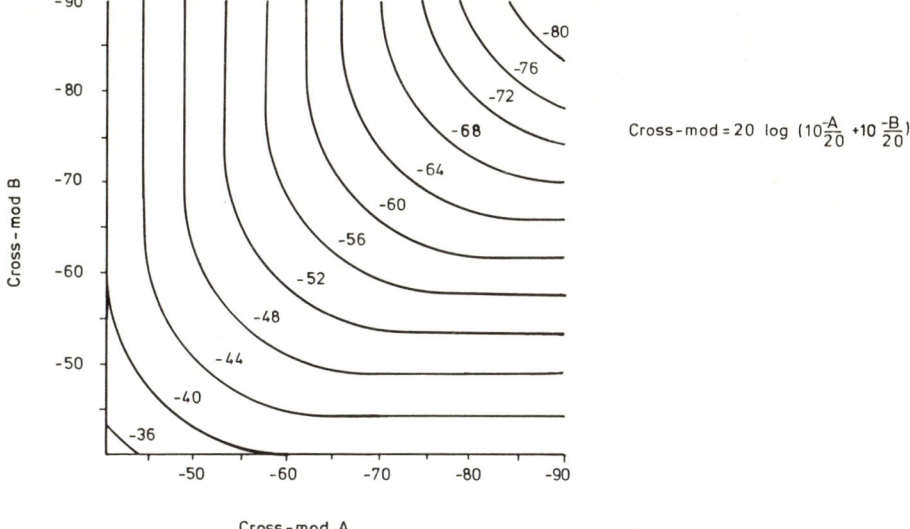

Fig. 8.8 Combining cross-modulation.

cascades, the reverse and the forward. Apart from the fact that the reverse amplifiers will be operating at very different gain and levels to the forward amplifier, it is still the case that the reverse amplifiers are not usually of the same quality as the forward units. This is because, in a long CATV system, every second reverse amplifier is often omitted, as the levels are high enough anyway, and the reverse frequency spectrum is just connected through via a jumper lead in the amplifier housing.

The equations are again quoted without proof. For two different amplifiers:

(1) Calculate C/N and cross-modulation of amplifier A.
(2) Calculate C/N and cross-modulation of amplifier B.
(3) Combine the two C/N figures using power addition:

$$C/N = -10 \log \left(10^{-\left(\frac{C/N A}{10}\right)} + 10^{-\left(\frac{C/N B}{10}\right)}\right)$$

(See Fig. 8.7.)

(4) Combine the two cross-modulation figures:

$$\text{Cross-modulation} = 20 \log \left(10^{-\left(\frac{A}{20}\right)} + 10^{-\left(\frac{B}{20}\right)}\right)$$

(See Fig. 8.8.)

For a two-way data system, the above calculations apply, but the cascaded C/N and cross-modulation of the forward and reverse channels are used to provide the two figures to be combined.

A practical example: a sub-split data system is built with 5 Texscan Phoenican II amplifiers, with reverse amplifiers.

Forward amplifier gain = 25 dB.
Forward noise figure = 7 dB.
Forward input level = 20 dBmV.
Forward cross-modulation at 30 dBmV = -95 dB.

Reverse amplifier gain = 10 dB.
Reverse noise figure = 12 dB.
Reverse input level = 30 dBmV.
Reverse cross-modulation at 32 dBmV = -105 dB.

FORWARD PATH
C/N:

Noise output of amplifier = noise + gain + noise figure
$$= -57.4 + 25 + 7$$
$$= -25.5 \text{ dBmV}$$
Output level of carrier is + 45 dBmV, hence
C/N of single amplifier is 70.4 dB
C/N of forward cascade $= 70.4 - 10 \log (5)$
$$= 70.4 - 6.99$$
$$= 63.4 \text{ dB}$$

Cross-modulation:

Cross-modulation ratio of single unit = 95 dB at 30 dBmV.
Hence cross-modulation ratio of single unit operating 15 dB above this carrier level (+45 dBmV output) is 30 dB worse, or 65 dB. (Note that cross-modulation level is 45 dB higher, but carrier is also 15 dB higher.)
Cross-modulation of cascade $= -D + 20 \log n$
$$= -65 + 20 \log (5)$$
$$= 51 \text{ dB}$$
(Note: at this point, due to high system levels, cross-modulation on the forward path is already marginal.)

REVERSE PATH
C/N:

Noise output of amplifier = noise + gain + noise figure
$$= -57.4 + 10 + 12$$
$$= -35.4 \text{ dBmV}$$
Output level of carrier is +40 dBmV, hence

C/N of single amplifier is 75.4 dB
C/N of reverse cascade = 75.4 – 10 log (5)
= 68.4 dB

Cross-modulation:

Cross-modulation ratio of single unit = 105 dB at 32 dBmV.
Hence cross-modulation ratio of single unit operating at 40 dBmV
is 16 dB worse or 89 dB.
Cross-modulation of reverse cascade = –89 + 20 log (5)
= 75 dB

COMBINATION TO FIND THE SYSTEM PARAMETERS

C/N:

We combine the forward C/N of 63.4 dB and the reverse of 68.4 dB.
From formula:

Combined C/N = 62 dB

Cross-modulation:

Cross-modulation to be combined are 51 dB for the forward path
and 75 dB for the reverse. From formula:

Combined cross-modulation = –50 dB

(Note: this example shows the danger of running amplifiers at too
high an output level – the reverse amplifiers, running 5 dB less level,
are much more comfortably within the specification.)

Table 8.2 is an aid to any calculations needing logs to the base 10.
To find the log of 34, shift the decimal point to 3.4, and look up the log.
As the decimal point has been shifted, add one to the log (i.e. log (34)
= 1.5314789. The number of times the decimal place is shifted is the
number that must be added to the log – the log of 340 is thus
2.5314789.

Table 8.2 Log table

	0	.2	.4	.6	.8
0		.698970	.397940	.221849	.096910
1	0	.0791812	.1461280	.2041200	.2552725
2	.3010300	.3424227	.3802112	.4149733	.4471580
3	.4771213	.5051500	.5314789	.4149733	.5797836
4	.6020600	.6232493	.6434527	.6627578	.6812412
5	.6989700	.7160033	.7323938	.7481880	.7634280
6	.7781513	.7923917	.8061800	.8195439	.8325089
7	.8450980	.8573325	.8692317	.8808136	.8920946
8	.9030900	.9138139	.9242793	.9344985	.9444827
9	.9542425	.9637878	.9731279	.9822712	.9912261

Automatic gain control (AGC)

The reason for incorporating automatic gain control in a CATV system is primarily to take care of attenuation variations caused by thermal effects. The amplifiers themselves are extremely stable and have no discernable gain variation with ageing, but the cable attenuation, as explained earlier, will vary by about 1% for every 5°C temperature change. AGC systems in CATV amplifiers usually have a gain control range of ±3 dB, which does not seem much, but will take care of the most violent temperature swings. On a trunk system, spaced at 22 dB, 1% attenuation change is only 0.22 dB, so to be noticeable, changes greater than about 40°C are required, producing about 1.5 dB change in the attenuation.

It can be seen that thermal effects are easily taken care of by AGC systems, and that such systems are only really needed on long, pole mounted outdoor runs. (A buried cable has practically no temperature variation.) The automatic gain control is, therefore, more of a thermal compensator than a true gain control. In practice, the gain of the amplifier is adjusted manually to the correct level, and then the AGC circuit switched in.

The usual method of providing AGC on a CATV system is to provide so-called 'pilot carriers' – a narrow band, pure carrier which is positioned at either end of the forward frequency spectrum. The AGC circuitry in each amplifier monitors the level of the pilot carriers and adjusts the amplifier accordingly. This technique leads to inexpensive ways in which to monitor the performance of the forward cable system, by doing rather more sophisticated monitoring of the pilot carriers, checking for absolute failure at critical system nodes, etc. AGC is almost never used on the reverse channels, due to the fact that it is difficult to locate pilot carrier generators. These, to be effective, must be at the end of every spur on the system. Since the cable attenuation is less on the reverse channels anyway, the absolute change with temperature of the cable attenuation on the reverse channels will also be less.

It should be noted that some CATV equipment, especially devices such as headend channel amplifiers, can have a form of AGC fitted, which is designed to give the device a constant output level from a varying input signal – such a unit might be used in a headend system processing marginal TV pictures. These units use the relative signal level as the AGC reference, and cannot be used with data signals, where the carriers are being switched on and off rapidly.

It is rare to use any form of AGC on an in-house data broadband network.

CHAPTER 9

Test Equipment for Broadband

This chapter is designed to provide a practical insight into the major items of test equipment and their use. The brief overview of the instruments given here is not meant to constitute an instruction manual for them, and the appropriate manuals should be consulted.

THE LEVEL METER

The level meter is one of the simplest, yet most important pieces of broadband test equipment. It is basically a wideband, tuneable radio receiver, covering the entire cable spectrum. It is invariably built as a portable device, and designed to measure the level of a television carrier directly in dBmV.

Basically, the unit is tuned to the frequency of interest, the wanted carrier found, and the level read off the meter. In practice, the use is slightly more difficult, as a sharp data carrier can be easily missed (the tuning of the devices is not too sharp) and, once found, must be 'peaked' correctly. The first reading will almost certainly be incorrect, due to the unit not being 'spot on' the correct frequency, and the tuning control should be rocked backwards and forwards through the wanted signal until the maximum is found.

As the meters are built for television signals, the receivers are quite 'wide' and a data system with closely packed carriers can be very difficult to measure properly. It is better, therefore, to use pilot carriers, or other test carriers away from the data.

As with all portable equipment, the internal batteries should be recharged regularly, and checked frequently in the field.

THE SPECTRUM ANALYSER

The spectrum analyser is expensive, and valuable as the 'ultimate weapon' for setting up and maintaining cable systems. Strictly speaking, it is not essential, as the same functions are performed by the tuneable level meter, but the time saving and ease of use are great advantages.

A spectrum analyser displays the frequency spectrum, over a given range, in the same way as an oscilloscope displays time over a given range. It consists of a tuneable receiver, automatically tuning from one end of its range to another. The received signal at any one instant, and therefore frequency, is displayed on a CRT. Looking at the display, one has an instant picture of the frequency spectrum, what signals are present, if noise is about, and what sort of noise, at what frequencies. Thus, at a glance, the spectrum analyser displays information that could only be very laboriously gathered with a level meter.

Modern units can provide quite accurate carrier level readings, if used properly. This is an important point: it is quite possible to make erroneous readings, by using the unit with too fast a sweep rate. On a data system, with sharp carriers, if the spectrum analyser's receiver is tuned through these carriers too fast, then it will not have time to respond properly, and the levels will appear too low. If there is any doubt about measured levels on an analyser, then the sweep rate should be reduced, and the amplitude of the levels rechecked.

THE SWEEP GENERATOR

The sweep generator provides an automatically varying frequency signal source with which to test the frequency characteristics of a cable network. The generator is set to vary slowly the frequency of the signal over the entire cable range. Viewed on a spectrum analyser, this signal can be seen, and the level variations caused by components noted. Invariably, sweep generators can supply fixed frequency signals as well, making them useful in level measurements of active systems. (It is not really possible to sweep a carrier through an active data system, as this would interfere with the data too much. This technique is used, however, with a fast sweep, on live television systems, as the picture degradation is momentary, and usually goes unnoticed.)

CHAPTER 10

Line Power of Amplifiers

The conventional, and in many cases only, method of powering amplifiers is via the cable itself. This is essential in CATV systems, where the amplifiers may be sited almost anywhere, but is rather a nuisance in data systems, where the amplifiers can be sited close to standard mains outlets. The systems used employ a low voltage AC power on the main trunk cable itself. From modern supplies, this AC is a square wave, rather than a sinewave, in order to increase the amount of power on the cable for a given voltage swing. In general, American systems use either 30 V AC or 60 V AC, whereas European systems have standardised on 48 V AC. The most common form of power supply is, however, the 60 V unit.

Starting with the power supply itself, these units are more sophisticated than the pure requirements would suggest. They are simply required to produce 60 V at 5 A or 10 A. This is usually done with a constant voltage transformer, and elaborate short circuit protection will be built in. Lightning arrestors, transient and surge protectors will be built in as standard. High quality units have a 'soft' turn on, i.e. they bring the voltage up on the line slowly, to avoid inrush currents in the line amplifier power units.

Because of these features the power supplies tend to be a higher proportion of the cost of a small data system, with perhaps one amplifier, than is thought. Having produced the 60 V, the next step is to put it on the cable. This is done with a 'line power inserter'. This unit has a cable input and output, and a connection for the power leads. The inserter contains filters to ensure that the radio frequency signals on the cable do not travel down the power leads. In addition, the unit contains two fuses – one in each cable direction. By removing one of these fuses, it is possible to direct the power feed to the appropriate leg of the cable. (As a rule, a 5 A power supply can handle about six amplifiers.)

The amplifiers themselves contain the circuitry to remove the necessary power from the cable, and usually have a power directing plug, where, like the fuses in the inserter, the line power can be fed to the next cable leg, or terminated. The direction that the power comes from is also a function of the power plug. Inside the amplifier is then a normal power supply, usually providing 24 V DC for the electronics. Very often test points are provided for this voltage, as well as the line volts. High quality units can have diagnostic LEDs to indicate the status of the power lines. This is important, as the most common cause of amplifier failure is still power supply trouble. For this reason, although switched mode power supplies are available, linear are to be preferred.

A feature of long CATV cable runs is that the cable resistance (loop resistance) will become quite high, in the order of some ohms. This means that there will be a substantial voltage drop between the power supply unit and the amplifier, and that the 60 V may be dropped to about 45 V. To cope with this, many amplifier power units have tapped transformers, with 'high, medium and low' voltage tappings, that are set according to the voltage measured at the amplifier line test point.

Although the amount of power on the cable is restricted, it is perfectly feasible to power other equipment from the line. The standard tap does not pass power to the drop cables, but some taps are available to special order that will do this. A standard line power inserter can be used in the reverse direction, to extract power. Units are available that take the 60 V power and transform it back to 230 V AC, so that low consumption mains powered equipment can be plugged in. This only becomes realistic if the cable power is derived from a secure power source, and the device to be powered is critical, such as a fire alarm or surveillance camera.

CHAPTER 11

Cable Systems for Data Communications

DATA SYSTEMS

The design of CATV and data cable systems has been considered in some detail, but the types of data device, and their impact on the system design, have not yet been covered.

Three basic types of broadband data networks can be identified:

(1) Point-to-point.
(2) TDM with headend controller.
(3) Distributed network with remodulator.

Point-to-point network

The point-to-point network is the simplest form of broadband data network. As has been shown, due to the use of directional taps and splitters, it is not possible for a data modem on one tap drop to 'talk' directly with another (Fig. 11.1). All that such a modem can do is to transmit, on a frequency in the reverse spectrum, to the headend. At the headend another modem is placed, which is the exact counterpart of the remote unit, i.e. it receives the reverse channel and transmits on a forward frequency. Such a system is thus full duplex, having independent forward and reverse transmission.

The physical headend of the cable system is also the logical start of the data system. This arrangement is satisfactory for large computer systems, running many terminals from one computer room. The headend will thus be a number of splitters, so that many devices can be connected. In a point-to-point system it is not essential that all the headend devices are in the same physical location, as a 'headend cable' can be run that is logically and electrically part of the headend, but physically remote. Point-to-point data links provide dedicated, high speed data channels, using very little bandwidth on the cable system – a typical 9.6 kbps modem uses about 25 KHz of frequency space.

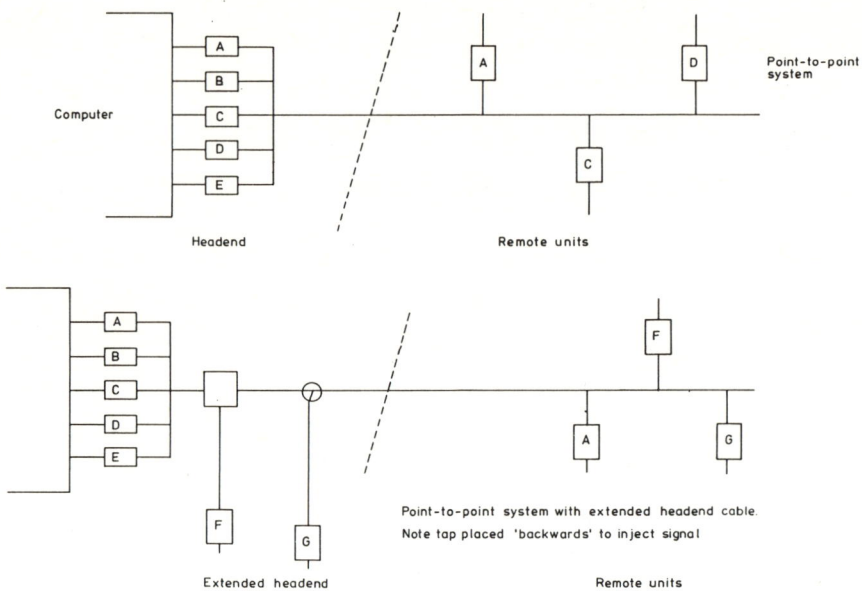

Fig. 11.1 Point-to-point modems.

(Time Division Multiplier) TDM with headend controller

TDM with headend controller (see Fig. 11.2) is used with a variety of data systems. Here, there is only one device at the cable system headend. As before, all the remote devices 'talk' back to the headend, but all units are on the same frequency. The controller polls the terminals or modems in turn, and any stored data is transmitted back to the controller at high speed (usually greater than 100 kbps). There are two advantages: first, there is only one device at the headend, and second, the cable system carries only one fairly narrowband signal – an extremely simple task. The rest of the cable frequency spectrum is kept free for other dedicated links, TV, etc.

Distributed network with remodulator

The third type of data system is the distributed network with a remodulator (Fig. 11.3). Up to now we have considered only data systems where, due to the directionality of taps, the modems communicated with the cable system headend. A true local area network system, however, demands that every device on the network shall communicate with every other device. In broadband cable

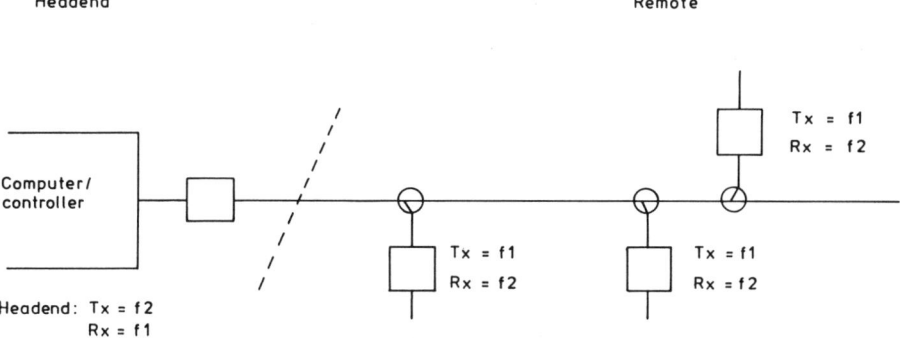

Either terminals or modems must be addressable, as the controller polls each in turn. Only one modem can transmit at any one time, although all receive all data from the headend. This scheme only uses one frequency pair on the cable, leaving most frequency space clear for other applications. Sometimes known as 'multi-drop' or 'multi-point' operation.

Fig. 11.2 TDM with headend controller.

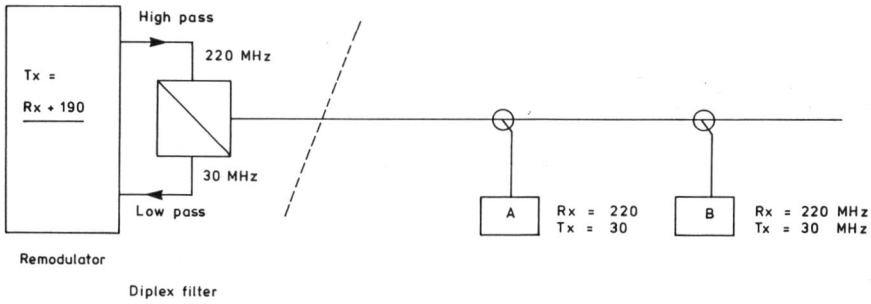

A remodulator usually operates on a block of frequencies (16, 18 or 36 MHz are standard) as shown below.

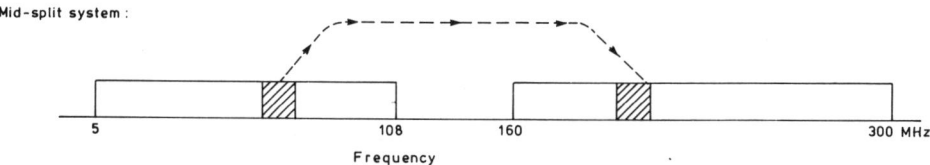

Fig. 11.3 Remodulated system.

terms, this means that every modem shall be able to receive a signal from every modem transmitter. The remote modem has a transmitter on a frequency in the reverse channels, and a receiver on a forward frequency.

To overcome this problem, a remodulator (sometimes called a 'frequency converter' or 'translator') is placed at the headend. This unit receives the reverse channel (usually a band of frequencies in the reverse channels) and retransmits it on a band of frequencies in the forward channels. A local network data system is made by tuning all the modem transmitters to the same frequency, in the reverse spectrum, and all the receivers to a forward frequency. The remodulator simply retransmits the received signal on that forward frequency. Thus all modems can receive the signals from each other. Obviously, such a network can only form half-duplex connections, and some form of network protocol, such as CSMA/CD or token passing must be used to form the distributed control.

It should be noted, in passing, that the choice of network protocol used is somewhat dependent on the size of the cable network. CSMA/CD techniques become inefficient if the propagation delay of the transmission media is high, as the number of collisions will increase, due to devices spaced widely apart transmitting nearly simultaneously, having first monitored the cable for carriers and found none, as the carrier from the station at the other end of the cable had not yet arrived. This is compounded in broadband, where the signals must travel to the headend, be remodulated, and then travel back along the cable. Broadband cable has a delay of some 5 μs/km, and a remodulator unit about 2 μs, due to the filtering. In a typical plant network, the length of the longest cable run is not very great, and this would not be a problem, but in some specialised applications, a token passing protocol would provide better response times.

Almost any topology except a ring can be implemented in a broadband data system – a broadband 'star' network is perfectly feasible, and in some buildings the best way of distribution from a main trunk.

HEADEND DESIGN TECHNIQUES

The headend of a CATV system can be very complicated – as already shown, by use of a 'headend cable' the physical headend can be stretched to cover two or more locations. When designing a broadband system, the physical headend should be placed in an accessible place, taking into account any present or future need for dedicated

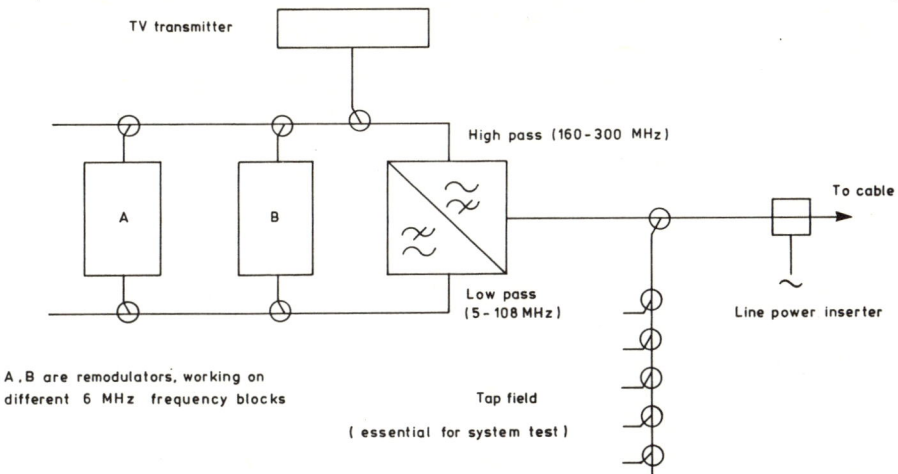

TV transmitter

High pass (160 - 300 MHz)

A   B

To cable

Low pass
(5 - 108 MHz)

Line power inserter

A , B are remodulators, working on
different 6 MHz frequency blocks

Tap field

(essential for system test)

Note: the losses in such a headend arrangement are such that it is common practice to follow the headend system directly with an amplifier.

11.4 Headend system design.

point-to-point links. Although point-to-point links can use a remodulator, it is wasteful of bandwidth (a full duplex link needs four channels), and if it is known that point-to-point links may be required, then the cable headend should be placed accordingly. Similarly, any plan to put television services on the network would affect the headend placing.

The headend cable layout should have two parts: a reverse tap field for point-to-point modems and test signal injection, and a diplex filter to allow connection of remodulators.

Taking the reverse tap field first, this consists of taps placed in the reverse direction, so that the signal from the modems or test units is injected onto the cable. Isolation at the headend is not usually a problem, so inexpensive multiple splitters may be used if the tap field is to be very large.

The diplex filter feeding the remodulator is not really essential, as the remodulator itself will also have these filters, but the cleaner the signal into the device, the better the output. If several remodulators are to be used, then the filter output will be fed to splitters. If video signals are to be inserted onto the network, they should also pass through the diplex filter, so that there is no reverse signal on the video transmitter outputs. A typical installation is shown in Fig. 11.4.

For a system with many differing signals on the cable the various channels can be separated, and spurious signals reduced by the use of individual channel filters. These filters are available covering single television channels, and are of very high quality.

CHAPTER 12

Introduction to Non-data Applications

One of the reasons for installing a broadband data system is that other services can be combined on the single cable. As the cable has been designed to the best TV transmission standards, it can be expected that TV will present no problems on a data network. This is correct for a well designed system, being run at the calculated levels.

In practice, the introduction of TV is not always trivial, as some system noise problems may show up. Data carriers are in general narrowband, and will produce narrowband second and third order distortion products. These may not show up on wideband noise measurements on the cable, but will produce picture interference. Television pictures are very sensitive to some forms of narrow carrier interference in the picture passband. Before placing the television channel on a data system, it is good practice to examine the proposed frequency space for such products.

The next problem, unique to the UK, is that there are no standard 625 line VHF television channels that can be received on a normal television receiver. This means that the television cannot just be plugged into the cable system as in the United States. Probably the best solution is to use a small VHF — UHF converter for each TV set. These are available at low cost, and provide good quality pictures.

At the headend, the TV transmitter unit is known as a *signal processor*, and can either take a direct video signal from a studio, or the standard vision intermediate frequency (IF) of 38.5 MHz. The signal is then modulated onto the wanted VHF TV channel. The practice is to receive, off air, the channels wanted for distribution, convert down to the standard video IF, and then to modulate up to the selected cable channels.

When running data and TV on the same cable it is usual to run the TV signal at a higher level than the data - 5–8dB is standard. This is to help overcome the narrowband distortion products discussed above. If it is known that the data system will have TV on it, the system should be designed at TV levels throughout, and the data system designed in at lower levels. Because of cross-modulation problems if amplifiers are driven at high signal levels, it may be necessary to reduce the amplifier levels, resulting in lower data signal output levels. The tradeoff between TV and data should be taken into account by the system designer.

To ease the task of headend design, the standard TV processor output signal level is +56 dBmV, so that the losses in the combiners and filters in the headend will not reduce the level too much. The standard output level for a television to function correctly is 0dBmV.

SECURITY TV SYSTEMS

In a security TV system, cameras are often located at extreme ends of a plant. The signals can be sent back to the headend on the reverse channels, but this is wasteful of bandwidth, and often of two-way cable capability. For a large number of channels, the best way is to extend the headend cable to the plant extremities and transmit directly on forward channels. (A good plan would be to use 440 MHz forward amplifiers and components, and place the television channels above 300 MHz, which would allow for about 15 UK standard channels.)

It should be noted that there is no 'correct' way to design such a combined data and security system – each case should be treated according to its individual requirements.

AUDIO

Audio on a CATV system takes two forms – high quality distribution of broadcast type (background music, paging), and telephone, including intercoms.

The former type of audio is quite easy to implement. There are standard FM transmitters for cable networks, and if a low split system is built, then conventional FM radios can be used to connect to the cable. At least fifty FM stereo channels can be accommodated in the 88–108 MHz band. On a mid-split system, standard FM frequencies cannot be used, but FM transmitters and receivers are available for other frequencies.

Telephone modems for cable use exist, and together with touch tone dialling, can be coupled to typical American PABX systems. The drawback to these systems is invariably cost, as the low volume of such devices leads to them being expensive. In general, it is not economically feasible to put an entire telephone system on cable. The main use of such telephone modems is to add telephone service to some remote plant location, such as a security gate, where the cable is used in connection with a security camera. It would be usual to power such security systems via the cable itself.

A feature of some current American public cable systems is that they carry telephone between two locations of a firm in the same town. This is usually done with a 'T1' carrier modem – a high speed modem carrying the American standard PCM telephone channels – 24 channels in one bit stream.

Whether this sort of application becomes possible depends on the public CATV licences issued in the UK. It is unlikely that the products exist to realise any particular audio application directly, and such applications will have to be handled by specialist hardware houses (e.g. for factory background music – this can be done with an FM radio receiver, but this must be placed in a secure enclosure, provided with power, etc.).

ALARMS

There are two types of alarm systems: those that provide warning of events external to the cable system, and those that provide an indication of the performance of the cable system itself.

External alarm systems are quite common in the USA and are made by many manufacturers. They invariably use the polling headend controller type of data system, where the remote modules are polled quite slowly, perhaps once every ten seconds or so. This allows a single, narrowband data channel to be used. A feature of such systems is that they usually use quite large computers at the headend, in order to cope with the stored database for many thousands of alarm contacts. There are some more simple units on the market, however, that allow polling by machines such as an Apple II microcomputer. Such systems are more applicable where the number of alarms is in the tens or low hundreds, as would be the case in a building security system.

The newer cable systems have modems built into the amplifier housings, to monitor and transmit the amplifier status, and to keep track of parameters such as power supply voltage, temperature, output level, etc. Once again, these units are polled in turn by a headend computer/controller.

Mainly because of the variety of different equipment to which an alarm unit must interface, there does not appear to be a 'standard' unit on the market.

CHAPTER 13

Cable Management and Certification

Broadband network management

Management of a broadband data network system is unlike that of any other form of data network. Traditionally, a data network performs a limited set of functions on a restricted data set format. A broadband network, due to the multiplicity of independent channels, will have many different types of communication - local area network operation on one frequency, point-to-point communications on another, television on another. Some of these services do not form a 'network' as such, but are simply broadcast services, using the cable transmission media. In the conventional sense of network management, i.e. monitoring of node points, a television distribution system is clearly almost impossible to 'manage' - moreover, the need to manage such a system in the computer network sense is not there.

A management system for a conventional computer network has two functions: first, the allocation and billing for the use of scare network media time - the use of expensive dedicated lines must be optimised; second, the recording of equipment malfunctions, so that failed modes can be bypassed, or other appropriate action taken. In the context of broadband networks, only the second function is really valid. The network media resource is no longer scarce, and will generally represent a small part of the total investment in computer power at a location. Although some companies would wish to make departments using a cable system pay on a usage basis, this would seem to be a false approach with what basically is a utility system - few departments will be charged for lighting or power on a direct usage basis.

This means that the prime function of a broadband network management system is to monitor the system performance. Due to the large number of independent networks running over the single physical cable, this is no trivial matter. There can be no single 'network controller' that can monitor the performance of every device on the system. To simplify the problem, it is realistic to divide it into two areas - individual network performance and cable system performance.

Individual network management must be implemented on a purely network basis. Many of the true local area network systems available for broadband, using forms of packet switching, have network management features built in. These units will be able to monitor the performance of that particular network, but will not be able to work with the other devices on the cable. Where possible, monitoring or management features should be built in, i.e. if a multidrop system is installed, then the terminals should respond to polling with a positive acknowledge type of protocol. It is essential that polled security systems are of this type.

For some types of system, such as pure point-to-point data connections, or broadcast channels, such as television, monitoring each device is not possible. The only possible form of network monitoring is to monitor the performance of the cable system itself. Amplifiers can now be monitored by devices in each amplifier housing, but there are at present no inexpensive ways to check the performance of passive cable runs. If such a spur on a network is really critical, then some positive acknowledge polled modem should be installed at the end of the spur. The device is not there to do anything except respond to polling, to indicate that the cable is good to that point. A typical data system will not have many long spurs, and almost invariably the spur ends will correspond to a security control location, such as a porter's lodge, and thus this scheme can be tied in with an overall security monitoring concept.

RF faults

RF faults can be characterised by loss of all or part service on the cable. This loss can be sudden, but some cable faults, such as sheath damage permitting water ingress, will develop over a period of time. Data corruption and signal level problems will result.

Cable damage will almost always affect the high frequencies first, so that devices which cannot receive the headend, but can be received, are an indication of cable problems. It should be noted that most cable television components are very reliable – trunk amplifiers have quoted MTBF of over 200000 hours.

Loss of service to a complete section of cable after an amplifier will usually be found to be caused by a line power failure. The power supplies in the amplifier units are the least reliable part of the system.

Modem faults

Faults attributable to modems depend on the nature of the modem itself: a packet switching device will show different faults from a point-to-point system. Isolated faults are almost always due to modem failure, but it should be remembered that all links have two ends, and it is not always the terminal end that is faulty.

BROADBAND NETWORK MAINTAINANCE

At the heart of broadband network maintainance is record keeping. The most useful tool is a plan of the entire system, with the locations of all fixed users (such as security alarms, etc.) marked. This plan should have the original design system levels marked up at critical points, such as amplifier inputs and outputs, splitter outputs and spur ends. This plan will enable speedy identification of faults on particular cable sections.

It is good practice on a large system periodically to check some levels on the system. This calls for the use of a signal level meter, to check the forward signal levels at the end of all the main cable spurs. (If the forward levels are correct, then it can be assumed that the cable itself is in order.)

It is a good plan to keep some form of plant record and take annual measurements. This will show up any form of long term deterioration of the cable network, and allow for any system additions that may have been made during the previous year.

Additions to a cable are made in exactly the same way as the design and installation of a new system, except that the levels are calculated from the point on the system where the addition is to be made. One of the reasons for using 22 dB amplifier spacing in a data network is that it allows insertion of a 3.9 dB trunk splitter, without exceeding the gain of the amplifiers. When addition of a new spur is completed, all downstream trunk levels should be measured and, if necessary, the amplifiers reset.

The security of a cable system can be maintained by the use of cable 'sniffers' – highly sensitive receivers designed to check for radio frequency leakage. They usually operate on one forward pilot frequency, and are used by walking along the cable, looking for an indication. Typically, poorly assembled connectors, badly closed amplifier housings, and missing terminations will cause leakage. Leakage can sometimes be detected by the reverse effect – the presence of noise, in the form of radio signals, on the cable. In particular, local, high power VHF FM transmitters will enter through missing terminations.

TESTING NEWLY INSTALLED SYSTEMS

(1) The exact procedure depends to a great degree on the available equipment. A sweep tester with return loss bridge (e.g. Texscan 9900) can greatly ease the finding of damaged cable sections.

(2) Install cable, taps and amplifiers. Leave out all power links, fuses, etc.

(3) Check cable section to first amplifier for ohmic short circuits. Sweep cable section if equipment is available – this shows up incorrect terminations, cable kinks, etc.

(4) Set up sweep generator at headend, and monitor input to first amplifier with either (a) a sweep display system, or (b) a spectrum analyser on slow sweep.

(5) If input is correct, proceed to amplifier set-up. If input is incorrect, work along cable with the display system, until fault is found. Set up amplifier forward channel to system calculations.

(6) Repeat procedure for all other amplifiers on forward channels.

(7) With the forward channels correct, it can be assumed that the cable is in order, as kinks, damage, etc., show up more in the HF end of the spectrum.

(8) The reverse amplifiers are set up using a sweep generator at the end of each amplified spur.

COMMISSIONING OF INSTALLED SYSTEMS

(1) After last amplifier, measure and record C/N of forward system.

(2) measure and record C/N of reverse system, using carriers at the end of each amplified spur.

(3) For a remodulated system (e.g. LocalNet), compute total system C/N.

(4) Measure and record all amplifier gain settings.

(5) Measure and record all critical location system levels on a frequency above the highest forward channel in use. These locations will be the end of all system spurs.

(6) The commissioning documentation to be held both on site and at the network vendors.

CHAPTER 14

LocalNet* A Broadband Local Area Data Communication Network

INTRODUCTION

LocalNet is an example of an integrated, local area data communication network serving not only future intelligent workstations, but also the substantial installed base of interactive terminals, time-shared computer systems, and minicomputers which comprise today's data networks. Integrating the technologies of wideband broadcast local networks, packet switching and cable television (CATV), LocalNet provides services inexpensively enough for today's needs but with the flexibility required for future applications. The generality of the LocalNet system enables it to serve efficiently a wide range of user applications, among them timesharing networks, management information systems, and integrated office communication systems. The particular advantages of the broadband communication system further extend its usage to industrial control and monitoring systems, as well as inter-campus and intra-city data distribution systems. Broadband CATV is unique in its capacity to integrate a wide variety of data communication applications, as well as other services, onto a single medium.

This chapter describes the LocalNet 20 packet communication unit (PCU) that connects one or more terminals or (local) devices to the broadband network, and through the network to other remote devices, also attached to the network through PCUs. A compact unit, an individual PCU is connected to coaxial cable routed within the walls of a building. The coaxial cable carries data within a building, between the separate buildings of a large organisation, and even over longer distances. The PCU, connected between the local terminal and the coaxial cable network, allows data communication over the LocalNet system with many other remote terminals, printers and computers. As a result, the local device can choose any network member as a correspondent device. The local device can also communicate with members of other LocalNets if the network has bridges and/or gateways to those networks.

* LocalNet is a registered trademark of Sytek Inc.

Technological background

Broadband coaxial cable (CATV) systems provide superior band-width characteristics, excellent noise immunity, and offer straight-forward user connection by inserting isolating taps on a single line or drop. The broadband system looks like a common bus to users at any physical access point. Standard frequency modulation methods are used on radio frequency (RF) carriers throughout the LocalNet portion of the RF spectrum, creating many independent LocalNet digital information channels through frequency division multiplex-ing (FDM). These LocalNet channels coexist with other, non-LocalNet, services of the broadband network. Examples of such services might include point-to-point broadband modems, broadcast and conference video, and specialised data services, such as energy management and physical security.

LocalNet SYSTEM 20 FUNCTIONS

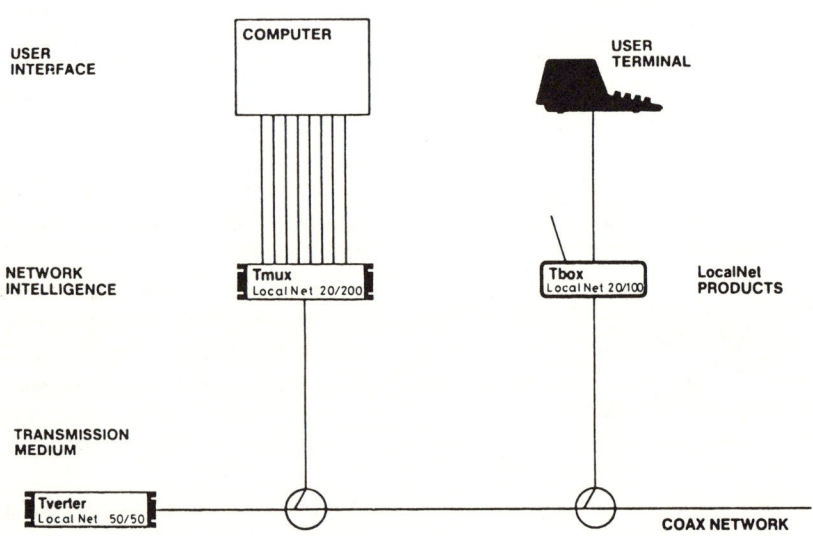

Tbox, Tmux are packet communication units. Tverter is a frequency translator.

Fig. 14.1 The intelligent network.

LocalNet time division multiplexes (TDM) these digital channels among a large number of user devices through the use of digital packet switching. Packet switching offers several advantages over other network architectures. Most significant are higher channel utilisation and optimised throughput provided by demand assignment of channel capacity to network users. Packet switching supports substantially greater numbers of data communications network users than can be handled by the more inefficient circuit switching techniques. Packet switching protocols provide efficient methods for user-to-user flow and error control, while providing the flexibility for user and equipment relocation and upgrade. LocalNet uses packet switching to provide time division multiplexed access to each FDM channel by many PCUs.

The intelligent network

LocalNet provides data communications intelligence in each CPU. This intelligence performs packet assembly, disassembly, buffering, error and flow control. These services allow the construction of packetswitched connections between correspondent PCUs, and sessions between their attached user devices. This intelligence can also economically provide the value added services of protocol and code conversion, speed matching, data security and more. The LocalNet network assumes responsibility for most data communications functions, from high level, application dependent requirements through complete transmission management, including network monitoring and other services. LocalNet's intelligent network is illustrated in Fig. 14.1.

Because network intelligence is contained within each LocalNet packet communication unit, control and reliability problems are minimised. Moving the network intelligence required for data communication protocols and functions into the network has the following advantages. This movement:

- Decreases network development and installation time by reducing the need to develop new user device software or hardware.
- Substantially increases throughput by using the LocalNet packet communication unit to send user data through the network.
- Increases the processing capacity of devices by offloading substantial portions of network communication processing to the PCU.
- Simplifies network growth by removing network dependent functions from user devices.

NETWORK ARCHITECTURE

LocalNet is a packet switched local area data communication network providing communication functions via a broadband CATV data distribution system. The properties of broadband CATV permit LocalNet to construct many independent subnetworks - termed *channels* - on one or more CATV coaxial cable systems. Each of these subnetworks provides data communication for hundreds of user devices.

User devices are attached to the network via an intelligent interface device called a *packet communication unit* (PCU). Special network services, particularly support for network management, are provided by LocalNet server nodes. LocalNet subnetworks (and associated PCUs, servers and user devices) can be connected through the use of packet switched bridges and gateways.

A typical LocalNet system is composed of a series of components:

- Packet communication units.
- Bridges and gateways for channel and network interconnection.
- Network management and control devices.

COMMUNICATION SERVICES

A LocalNet 20 PCU provides a variable number of full duplex sessions based on switched virtual transport connections to other PCUs in the LocalNet. Sessions can be formed to another port on the same PCU, between PCUs on the same LocalNet channel, or between PCUs on different channels and cable systems via one or more interchannel bridges or gateways. Flow control is maintained between PCUs in order to match the speeds of user devices participating in a session - pacing each transmitter to the speed of the receiver, compensating for heterogeneous bandwidth, delay and error properties of the network, including statistical internal network congestion.

There are two user device addressing forms: a fully qualified address specifying a particular device port on a particular PCU, and a partially qualified rotary address specifying any available port on a particular PCU or group of PCUs.

The user device interface to the PCU (and thus to the LocalNet) is controllable either locally by the user device or remotely by the network manager. This supports overall LocalNet network control, permitting remote monitoring and control of a PCU from a network control centre by the network manager.

BROADBAND COAXIAL CABLE

LocalNet uses broadband coaxial cable as its physical communication medium. In wide use in many commercial CATV applications, this medium provides the high bandwidth (300–400 MHz), proven reliability and multidrop capability required for growing data communications requirements. Broadband cable provides high noise immunity, with each user branch electrically isolated from the main signal. This isolation also reduces the possibility of system failure from accidental or malicious causes. Analogue video or voice applications can share the same cable using dedicated frequency channels.

Broadband CATV is a directional broadcast transmission system based on a single inverted tree physical topology. Each user device can broadcast its transmission in one direction, up the tree towards the CATV headend. The headend contains an analogue frequency broadband translator (LocalNet 50/50) that, like a satellite transponder, rebroadcasts the transmission from the headend of the tree, forward to all attached user devices.

LocalNet exploits the directional aspect of the CATV transmission system to achieve full connectivity, and provides products that are

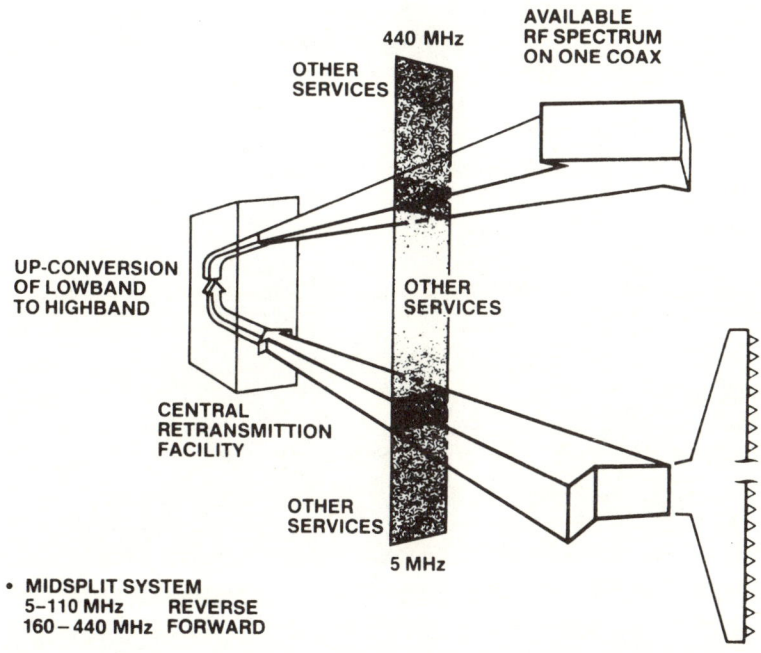

Fig. 14.2 Bidirectional mid-split broadband channel organisation.

compatible with each of the major cable system architectures: mid-split, sub-split and dual-cable. Bidirectional transmission on a mid-split LocalNet is illustrated in Fig. 14.2.

For LocalNet usage, the 300 MHz cable bandwidth is divided into multiple logical channels through the use of frequency division multiplexing (FDM). Frequency shift keying (FSK) is used to modulate individual RF carriers at a 300 KHz spacing to create LocalNet 20 data channels, with each channel accessed by multiple PCUs spread over the geographic radius supported by the broadband transmission system – up to 50 km.

DATA CHANNELS

LocalNet 20 supports up to twenty FDM data channels with data rates of 128 kbps, and provides packet switched data communications service to thousands of low duty cycle devices, such as terminals.

A simple type F coaxial cable fitting in each area to be served provides access to the underlying analogue broadband network, and hence to LocalNet 20 service. The independence of the transmission system from the types of services offered makes connecting, disconnecting and moving the PCU from room to room a simple process, with no wiring changes required. Functional upgrades or the addition of new services, such as voice and video teleconferencing,

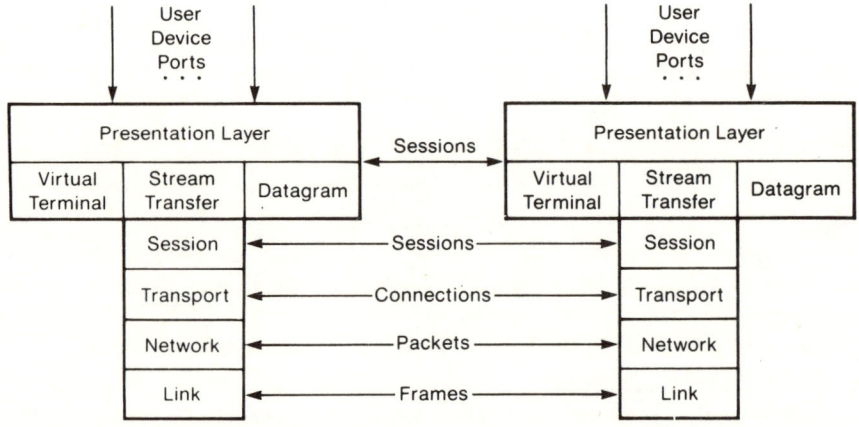

Fig. 14.3 LocalNet protocol architecture.

may be made without addition or change to the analogue CATV transmission system.

COMMUNICATION PROTOCOLS

Transmission bandwidth on a channel is allocated in a distributed manner using a carrier sense multiple access with collision detection (CSMA/CD) access method. This technique permits each packet communication unit to contend independently for transmission access with other PCUs on the same channel. The minimisation of centralised network control provides for high system reliability, low transmission delay and high channel utilisation. The LocalNet protocol architecture is illustrated in Fig. 14.3.

This architecture allows user devices to communicate primarily through *sessions*. User initiated sessions provide the communication services to each port. Many sessions can be attached to (and controlled from) each port, and many ports can be attached to (and controlled from) each user device. In general, each transaction between host and PCU is labelled, either directly or indirectly, with the port and session numbers to which it applies.

LocalNet employs a layered architecture conforming to the Reference Model for Open Systems Interconnection (OSI), developed by the International Standards Organisation (ISO). This model specifies a seven layer architecture that facilities communication among open, or standardised, systems. The seven layers are described below:

Physical layer. The physical layer within LocalNet is the broadband coaxial cable information distribution system.

Link layer. LocalNet uses several link layer protocols. The most important of these provides for channel level addressing, packet framing, and cyclic redundancy check (CRC) generation and checking, as well as channel bandwidth allocation using distributed CSMA/CD control. Other link protocols are used, as required, in bridges and gateways.

Network layer. LocalNet implements a datagram network layer that provides datagram-based addressing and routing of packets.

Transport layer. LocalNet provides several transport layer services, the most important of which is flow- and error-controlled connections between PCUs.

Session layer. LocalNet PCUs implement session layer services that provide access control, service accounting, name directory and extended addressing services.

Presentation layer. The presentation layer provides data formatting and conversion functions. LocalNet PCUs support presentation layer functions.

Application layer. The application layer provides application specific services to user devices. Generally, implementation of functions in this layer is left to the user.

The physical layer

Broadband coaxial cable is the physical foundation of LocalNet. LocalNet divides the available bandwidth of the coaxial cable into a number of frequency division multiplexed channels which are allocated for low speed (LocalNet 20) applications, as required. LocalNet PCUs transmit on the cable via radio frequency (RF) modems that convert the digital data stream to an FSK modulated analogue signal. As illustrated in Fig. 14.4, LocalNet modems are an integral part of each PCU.

The cable system is passive and does not impose a specific digital protocol on user devices and PCUs. The modems are packaged with the PCU, making the broadband cable system independent of the application, capable of supporting simplified service upgrades and extensions without cable system changes.

The connection of a LocalNet device to the coaxial cable and to the cable facilities is passive; connection or disconnection of the LocalNet device, or of a user device, does not affect the operation of either LocalNet or the cable system. LocalNet transmissions are contained in two frequency bands: an inbound band (low frequency) and an outbound band (high frequency). The inbound and outbound directions are usually referred to as the *reverse* and *forward* directions, respectively. Each modem transmits at a selected low frequency and receives at a correspondingly higher frequency. An

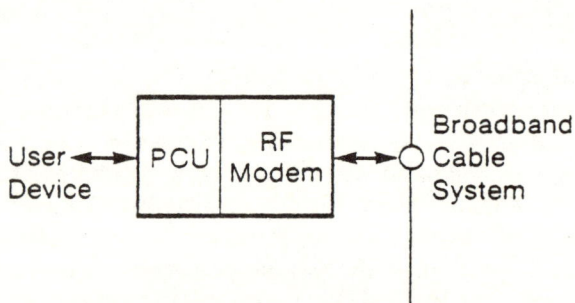

Fig. 14.4 User device attachment.

analogue frequency translator, the LocalNet 50/50, joins corresponding low and high frequencies to construct logical LocalNet digital channels.

Channels are allocated in groups of 20, and correspond to standard 6 MHz commercial video channel bandwidths. Transmission and channel allocation methods are RF analogue.

The data link layer

The data link layer transfers user data and commands between network nodes by way of the coaxial cable medium. The unit of information exchanged at this layer is referred to as a *frame*; frames are encoded in a high level data link control (HDLC) format, which has been modified to support better the local networking environment. HDLC allows for variable data record lengths through the use of flag characters to delimit the frame. Data integrity is assured through the use of a frame check sequence (FCS) algorithm which provides error detection and enables error correction through retransmission by the transport layer protocol. Automatic zero insertion and deletion allows for the transparent transmission of user data.

The use of an HDLC-derived data link protocol, coupled with the CSMA/CD cable-access technique, provides for the fair and equitable allocation of channel bandwidth to all LocalNet users, without requiring a centralised control mechanism.

The network layer

The network layer provides an unacknowledged packet transfer service between packet communication units on the same channel, on differing channels, or on differing LocalNet cable systems. Internal routing of packets between LocalNet elements is the fundamental function of the network layer protocol.

The transport layer

The transport layer provides end-to-end data transport via a reliable stream protocol (RSP) that is optimised for high throughput, providing reliable, flow controlled transfer of user data via a virtual connection. Up to four simultaneous virtual connections can be supported by a LocalNet 20/100 PCU. The LocalNet 20/200 PCU supports up to sixteen simultaneous virtual connections. One or more virtual connections can be associated with each user port. High data integrity, as well as the in-sequence delivery of user data, is obtained by the use of sequence numbers, end-to-end positive acknowledgements, and packet retransmission. Control over the flow of data across the virtual connection and to user devices is provided by a windowing algorithm.

The session layer

The session layer provides the network and session management functions to LocalNet packet communication units. Session layer protocols are responsible for resource protection, data security, network authentication, and symbolic name translation.

The presentation layer

The presentation layer provides data communication through the virtual terminal service. This consists of a single transport layer virtual connection constructed from network layer datagram packets.

The virtual terminal service is oriented towards the support of many low speed devices transmitting essentially character-at-a-time traffic. It provides a host interface that maximises host terminal buffer utilisation and provides tools for controlling the format of data presented to and received from remote terminals attached to LocalNet 20 PCUs.

The applications layer

LocalNet packet communication units provide a common, high performance, high function computer network that minimises the effort needed by users to implement sophisticated applications to the user's devices.

NETWORK MANAGEMENT FUNCTIONS

The OSI Reference Model provides for the implementation of network management protocols and services. These services allow for the configuration, test, control and management of an operational network. The following network management services are implemented in LocalNet 20:

• Diagnostic and maintenance facilities.

- Network startup and shutdown.
- Performance monitoring.
- Reconfiguration.
- Network security.

PACKET COMMUNICATION UNITS

LocalNet provides communication intelligence at each user interface point implementing applicable protocols from the LocalNet protocol suite. Services typically provided include: packet assembly and disassembly, buffering, session management, error and flow control, protocol and code conversion, speed matching, and data security.

NETWORK INTERCONNECTION

LocalNet provides three methods by which packet communication units (and their associated user devices) can communicate. First, for users connected to PCUs configured on the same LocalNet channel, the basic protocol provides complete one-hop connectivity. Second, for users attached to different channels on the same cable, an optional store-and-forward packet mode *bridge* routes packets between LocalNet channels. Third, for users attached to different cable systems, each supporting LocalNet, the *gateway*, a variant of the bridge, provides for interconnection of these distinct LocalNet systems into one common network through use of one or more transit networks: Digital Data Service, a public packet switched value added network (VAN), or other data communication system.

LocalNet gateways provide data communication between LocalNet installations and external networks, including X.25 packet networks. These gateway products both extend the size and geographical coverage of LocalNets, as well as supporting their interconnection with other types of networks.

SERVICES AND OPERATION

This section introduces the basic concept of operation of the LocalNet 20 packet communication unit (PCU), and defines each of the PCU's features and attributes. It concludes with a basic overview of the internal and external construction of the LocalNet 20 PCU.

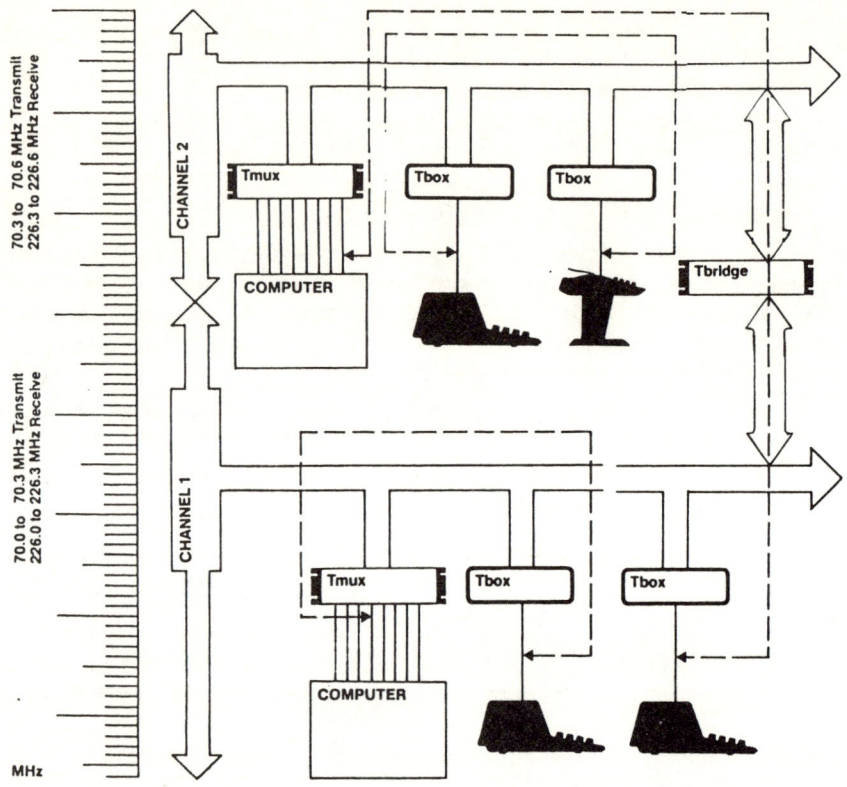

Fig. 14.5 LocalNet System 20 logical layout with interconnections.

Concept of operations

The LocalNet 20 PCU is designed as a network interface for serial interface devices, such as asynchronous terminals and computer systems. The LocalNet 20 PCU uses a common packet switched backbone network, permitting communication with other devices interfaced to the network through LocalNet products.

The essential service provided to user devices by LocalNet 20 PCUs is the full duplex transport of data bytes between correspondent user devices over packet switched *sessions*. A LocalNet 20 PCU interfaces its user devices via *ports* that provide access points to a virtual terminal service provided by each PCU. Sessions provide the communication medium between ports. A total of four sessions can be supported concurrently by each LocalNet 20/100 PCU, apportioned in any manner to the two device ports.

A LocalNet 20/200 PCU can interface as many as eight user

devices. A total of sixteen sessions can be supported concurrently by each 20/200 PCU, and these sessions can be apportioned in any manner to the eight 20/200 user device ports.

The PCU provides the means for user devices to initiate sessions to other devices attached to the same or a different PCU. The requesting user device need not be aware of the physical or logical location of its correspondent, as the LocalNet system provides all the necessary routing services. A source PCU, once a session has been initiated, takes data provided by a transmitting user device, formats that data into packets and uses the LocalNet internal data communication protocols to route that packet to the destination PCU. The destination PCU removes the data from received packets and transmits the data to the receiving user device. Data transmission is full duplex so that both correspondent user devices can be simultaneously transmitting and receiving. The manner in which data is presented to the PCU, and by the PCU to its attached devices, is under the control of the device.

LocalNet 20 channels are organised into six frequency groups. Each PCU contains a frequency agile RF modem capable of directly tuning to the channels in its frequency group. PCUs with modems belonging to the same frequency group can communicate by selecting the same channel. PCUs with modems belonging to differing frequency groups, or prohibited from changing their modem channel, can communicate only through an intermediary, LocalNet's interchannel packet mode store-and-forward *bridge*, as shown in Fig. 14.5.

Communication services

Each LocalNet 20 PCU provides a range of communication services to its attached devices. These services include:

- *PCU management*. Overall management of the PCU, including specifying its identity, location and protection attributes.
- *Session management*. Management of individual sessions, including session initiation, termination and control.
- *Data handling*. Handling and processing of user data through sessions, including flow-and-error control.
- *Signalling*. Out-of-band signalling across session boundaries.
- *Maintenance*. Facilities provided for network diagnostics and maintenance, particularly local PCU diagnostics.

The PCU features described above are controlled by a set of attributes set via PCU commands from the device. Attributes are divided into two categories: those maintained for the entire PCU, and those maintained for each PCU port. Sessions can individually control their per-port attributes.

CAPACITY
The 20/100 PCU supports a maximum of four simultaneous sessions and two user device ports. The 20/200 PCU supports a maximum of sixteen simultaneous sessions and eight user device ports.

PERFORMANCE
The throughput of a LocalNet 20 PCU depends on the characteristics of traffic between it and its user device. Sending single characters at low speeds requires more processing overhead than sending files at higher speeds. The aggregate throughput is of the order of 20 kbps. Some settings of the PCU data handling attributes can increase delay in processing a packet while also increasing throughput; see the discussion on intercharacter idle timer.

User device characteristics

LocalNet 20/100 and 20/200 PCUs interface asynchronous interactive terminals to the LocalNet. These devices can range in data rate from 75 to 19.2 kbps, and may use various types of parity and numbers of stop bits. The hardware interface is EIA RS-232C compatible. For user devices such as host computers and block mode terminals, where the aggregate input data rate to the PCU is greater than 2400 bps, flow control is recommended in order to prevent loss of data.

PCU/device interface

A PCU has three device interface modes: *command, autobaud* and *data transfer.*

COMMAND MODE
In command mode, a user device may input commands to either the local PCU or a remote PCU. When the PCU is reset, initially powered up, or has a terminal initially attached, the user interface is in command mode – indicated by the prompt.

If there are no sessions on a port, and the port is not in autobaud mode, it is in command mode. While in command mode, the PCU responds to flow control requests from the user device. However, the PCU does not request flow control of input from the user device.

AUTOBAUD MODE

A PCU enters autobaud mode after the rear panel reset button has been pushed twice within a few seconds. In autobaud mode, the PCU ascertains and matches the baud rate of devices attached to all its ports. The PCU searches the incoming data stream, beginning at a rate of 19.2 kbps. The device must supply ASCII carriage returns until the PCU has matched data rates with its attached devices.

In addition, each PCU will accept an RS232 interface signal that causes the port to autobaud. This autobaud can be accomplished at any time while Data Terminal Ready (DTR) is asserted.

Each device is placed in command mode after this reset. Receipt of the PCU command mode prompt announces a successful reset.

DATA TRANSFER MODE

Once a session has been established, by use of the *call* or *pcall* command, the user device enters data transfer mode from command mode automatically. At this point the user can log in to a remote device and begin sending and receiving data.

TRANSPARENT MODE

Devices often require the use of a connection on which all eight bits of each byte may be transmitted transparently, without interpretation by either the source or destination PCU. For this purpose, transparent mode is defined in the following way:

- *None* or *break* is specified as the command mode entry string.
- *None* is specified for the end of message delimiter (EOM).
- *None* or EIA is specified for source and destination local flow control.
- *None* is specified for parity.

When the specified attributes are set as shown, the PCU does not interpret characters in an attempt to derive command mode invocation, message delimiters, or flow control characters. In addition, the PCU does not interpret or modify the eighth bit of each byte as parity.

Global PCU management

PCU management attributes are of two types: those global to the PCU and all its ports and sessions simultaneously, and those specific to a particular port and its sessions.

Global attributes include: unitld, location (linkAddress, channelNo), group, privilege, and enabled commands. The value of these attributes is the same for all ports on a given PCU.

Global attributes control the *identity* of the PCU on the LocalNet, its *location* on the LocalNet, and its protection state.

ADDRESSING

To support information routing, each PCU possesses a set of address attributes, including:

- The PCU's unique *unitld*, used for PCU-to-PCU routing in the network layer and for session initiation.
- The modem group for which the PCU is configured.
- The modem *channelNo* to which the PCU modem has been tuned.
- The link access protocol *linkAddress*, used for on-channel link layer routing.

Generally, a remote PCU port is identified only by the PCU *unitld*, as well as required *portlds*. The *unitld* is specified for each PCU at its initial configuration time. It is a 16-bit value – represented as four hexadecimal digits – and permits over 65 000 PCUs to be configured onto a single LocalNet.

The modem group specifies which group of up to twenty LocalNet 20 channels the PCU's modem can directly tune. The modem group is determined by the hardware installed in each PCU and is specified at the factory. The modem group is indicated by an alphabetic value from A to F, and from L to N, inclusively.

A PCU's location is specified by its *channelNo* and *linkAddress*. The *channelNo* identifies which of the LocalNet 20 channels is to be used by the PCU for its communciations with other PCUs. Each PCU has the ability to make limited changes to its location by tuning its modem to one of twenty *channelNos* addressable within its modem frequency group.

A local user, if permitted, can alter his or her PCU *channelNo* within the range of the PCU modem group. The *linkAddress* provides an internal hardware address that is used to minimise internal processing overhead. Alterations to the *linkAddress* are made by the network manager.

A PCU's location, both *channelNo* and *linkAddress*, can be changed independently of its *unitld* while the PCU retains its membership in its LocalNet. PCU location changes in a large LocalNet are tracked by inter-channel bridges and the network control centre.

ACCESS CONTROL: COMMAND PROTECTION AND PRIVILEGE

LocalNet 20 PCUs provide tools to control access to the system facilities. This control is based on two facilities: enabled commands and privilege.

PCU global attributes are summarised in Table 14.1. Each attribute is retained in the non-volatile attribute memory; these attributes are also retained across power failures and normal PCU resets.

GLOBAL PCU MANAGEMENT COMMANDS

The PCU attributes are controlled by the following commands:

disable. Disables named commands from non-privileged invocation.

enable. Enables named commands for non-privileged invocation.

group. Specifies to the PCU its modem frequency group. Normally the group is set at the factory and the command is not further used.

location. Specifies to the PCU, the LocalNet 20 *channelNo* to which it is to tune its modem, and the *linkAddress* to be used on that channel.

privilege. Enables/disables the privilege state of the PCU.

unit. Specifies the *unitId* of the PCU.

These commands control the availability of commands for each PCU, its location and group numbers, and the ability of the PCU to control other PCUs remotely.

Session management

A *session* is a data communication connection between a local user device and a remote user device through the respective PCUs. Each session has its own set of attributes controlled by PCU commands. A session can be initiated or terminated from the local device, and the device can be switched among two or more sessions attached to its

Table 14.1 PCU attribute summary.

Attribute name	Possible values
unitId	0000-*FFFF* (hexadecimal)
linkAddress	0–254 (decimal)
channelNo	0–119 (decimal) with 300 kHz channel spacing
enabled commands	all commands
privilege	on or off
group	A–F (mid-split) L–N (sub-split)

port. In addition, sessions can be controlled remotely from the correspondent and third party PCUs.

ADDRESSING

A user-level address has two parts: *unitld* and *portld*. The *unitld* uniquely identifies the selected PCU on its LocalNet, and the *portld* identifies the requested access point at which a user device is attached to the PCU.

The address used to establish a session with a remote device can have one of the five following forms:

> *unitld, portld*. A fully qualified address.
> *unitld*. A partially qualified (sequential rotary) address.
> *unitld*. A partially qualified (random rotary) address.
> *unitld: rotarysize*. A random rotary address with a rotary size limit.
> *unitld: rotarysize, servicetype*. A random rotary address with rotary
>   size limit and service type identified.

A session initiation request that is fully qualified specifies the LocalNet access point (*portld*) to which the session request is directed. If the requested port is unavailable, the request is denied. If the partially qualified address is used (the *portld* is omitted), the PCU places a sequential *rotary* call. If the optional partially qualified address is used (the *portld* is omitted and the *unitld* is followed by a colon), the PCU places a random rotary call. The rotary size and service type parameters can be included in this random rotary call scheme.

Using a partially qualified address permits any of a group of equivalent points, called a *rotary group*, to serve as the destination of the session. This group consists of ports:

(1) which are on PCUs with *unitlds* within the range of the rotary;
(2) where the number of current sessions is less than the specified maximum number of sessions;
(3) which, if Data Terminal Ready (DTR) is not set to *off*, have the EIA pin DTR asserted by the destination device (e.g. computer port);
(4) that have *listen* set to *on*.

A rotary call is made to the first free port found within the PCU rotary group. The first unassigned *unitld* terminates the rotary. For example, the *unitld* sequence:

1234	1235	1236
1238	1239	
1241		
1243		

contains four rotary groups: 1234 through 1236, 1238 through 1239, 1241 and 1243. A rotary call to a group will result in a session with the first available port within that group. If no free port is found within the group, the session initiation attempt fails. The PCUs within a rotary are not confined to the same LocalNet channel if a bridge is used, nor do they have to serve the same user device.

The optional random rotary call scheme can be implemented to specify a range of unit calls placed randomly within the rotary group to distribute network traffic efficiently and skip over possible malfunctioning units.

SESSION MANAGEMENT COMMANDS
The following commands provide session control, as well as set the values of session management attributes:

call. The *call* command initiates fully qualified and rotary sessions.

done. Terminates either current or non-current sessions.

help. Displays a list of commands enabled for the requesting port.

listen. Enables/disables the port to respond to incoming session requests.

maxsession. Specifies the maximum number of sessions that can be attached to a given port.

sstat. The *sstat* (short status) command supplies the PCUs *unitld*, *portld*, and a list of established sessions for the port.

status. Supplies a summary of all global PCU and port-specific attributes, as well as a list of established sessions for the port.

suspend. Places a session into a flow controlled state.

switch. Changes the current session, re-activates it if it was suspended, and places the port in data mode.

SESSION INITIATION
A PCU responds to both local and remote requests to initiate sessions. A local session initiation request takes the form of a *call* command from a user device attached to the local PCU. The session can be established either to another port on the same PCU, or to a port on a remote PCU. A remote PCU can request the establishment of a session between two other PCUs.

Sessions can also be established using the permanent session mechanism (*pcall*) to call a preselected PCU (*punit*). Attempts to establish a permanent session can be initiated either automatically or by the occurrence of an external event. In automatic mode, a call is automatically initiated whenever the local port has no current session. In the *on* mode, the external event which triggers *pcall* can be either a carriage return entered by the user device when there are no sessions, or the raising of DTR by the device (if DTR control is not set to *off*).

FULLY OR PARTIALLY QUALIFIED ADDRESSES

The session request can use an address which is either fully or partially qualified. If the address is fully qualified, the request is for a specific port. If the port has a free session available, the request is granted and the session is established. If the address is partially qualified, the request is for a rotary and the local PCU responds by establishing a session to the first free port enabled for incoming sessions. If no such port is so enabled, the request is denied. Each session is labelled with a port specific *sessionNo* identifying the session. The *sessionNo* is assigned sequentially from 1 when multiple connections are established.

SESSION ESTABLISHMENT

A remote PCU cannot accept an unprivileged request for a session if the DTR control parameter is set to *on* and the DTR line of its EIA interface is not asserted, or if a DTR timeout is specified and DTR is not raised within the specified time. If the calling PCU has its PRIVILEGE attribute enabled (set to *on*), the remote PCU accepts a session request regardless of the DTR state or the DTR control parameter. If a PCU has been enabled to receive incoming sessions, ring indicator (RI) is raised when an incoming session request is detected. RI is lowered after 1 sec.

If DSR (Data Set Ready) control parameter is set to *on*, the PCU asserts DSR when it establishments its first session (after having no current sessions).

If DCD control is set to *on*, the PCU asserts Data Carrier Detect (DCD) for each port having an active session. If DCD control is set to *Auto*, the PCU asserts DCD for each port having two active sessions. If all sessions on that port are terminated, DCD is dropped.

SESSION TERMINATION

The PCU supports termination of sessions through the use of the *done* command. A local user device can terminate both the current session and any non-current sessions. The remote PCU may also terminate a session.

Sessions may be terminated if a timeout value has been specified for the local or remote PCU and if no character has been sent or received during the specified period. Also, if DTR control is set either to *on* or to a timeout value, and DTR is dropped by the user device, all outstanding sessions for that port are automatically terminated.

If DSR control parameter is set to *on* or $<x>$, the PCU drops DSR when the last session is terminated. When DSR control is set to $<x>$, DSR is reasserted after $<x>$ × 10 ms.

MULTIPLE SESSIONS

The 20/100 PCU supports up to four simultaneous sessions which may be allocated to a single port or distributed to both ports. The 20/200 PCU supports up to sixteen simultaneous sessions which may be allocated to a single port, or distributed among the eight user device ports.

The number of simultaneous sessions allowed for a particular port is determined by its maxsession parameter. For instance, on a 20/200 PCU, port 0 might have a maxsession limit of 8, while ports 1, 2 and 3 have limits of 5, and ports 4 through to 7 have limits of 4 each. If ports 2, 5 and 6 have all sessions connected, 13 of the 20/200 PCU's sessions are in use. Port 0 can now have no more than 3 sessions (making the 20/200 PCU's total of 16), even though the maxsession limit for port 0 is 8.

On a port, one active session is marked as the current session, and all input from the user device is directed to this session for transmission. On a status display, the current session is indicated by an asterisk after the unit and port numbers of the remote CPU. Sessions that are not current may remain active, or be *suspended*. An active session continues to receive and to deliver output data to the local device. To prevent data from another active session from being mixed at the user device with data and commands from the current session, a session may be suspended, thus flow controlling its connection, preventing subsequent data transport until reactivated.

SESSION SWITCHING AND SUSPENSION

A device can request its PCU to switch between sessions. The swtiched-from session remains active until explicitly suspended. If the switched-to session had been suspended, it is reactivated as a side effect of the switch.

If a device issues a new session request while any session is open, the completion of the new session request automatically switches the device port to the newly opened session. If a session is initiated to a remote port which already has a session, and the remote port is in command mode, the remote port is switched to the new current session. However, if the remote port is in data mode, it does not switch to the new session.

REMOTE COMMAND EXECUTION

Any enabled PCU command can be invoked either by the local device or by a remote PCU via an outstanding session. The status of remotely executed commands is routed back to the originating PCU port through the current session. Disabled commands are unavailable, except to a privileged PCU. Flow controlling a session inhibits receipt of remote commands.

Data handling

The user device may specify several attributes that control the manner in which its PCU port presents and processes data over the sessions attached to that port.

FLOW CONTROL
LocalNet 20 PCUs support the capability to control the flow of data to and from attached devices. Flow control provides for:

- Speed matching between devices of differing transmission speeds.
- Compensation for links of differing bandwidth, delay and error rate, between linked LocalNets.
- Reliable data delivery in the event of network or PCU congestion.

Each LocalNet 20 PCU responds to flow control signals both from its attached devices and from the remote PCUs with which each of these devices is communicating. Each session's underlying transport virtual connection propagates flow control information between correspondent PCUs.

LocalNet 20 supports two local flow control modes: EIA and xon/xoff. A device may select one of these modes to apply to its PCU port. The type of local flow control may differ between correspondent PCUs.

PACKET CONSTRUCTION AND TRANSMISSION
The PCU makes packets of the characters sent by a device for transmission to a remote PCU, and separates received data packets into characters for subsequent transmission to the local device. A device can define the conditions under which packets are transmitted to a remote PCU. Four conditions result in transmission:

(1) PCU input buffer full.
(2) Inter-character idle timer timeout.
(3) Logical end-of-message.
(4) An EOM count.

When the internal PCU input buffer is full, the contents of the buffer become the data portion of a packet and are transmitted. Packets are always transmitted when the buffer is full.

The PCU maintains an idle timer to measure the time elapsed between input characters. The timer clock is restarted after each character is received.

To ensure prompt delivery of single characters when using a full-duplex terminal, the buffer is closed and its contents are queued for packet transmission, if no subsequent character is input before the designated time expires. The user can specify the value of the idle timer in units of 10 ms.

LocalNet 20 PCUs support the definition of logical messages as sequences of characters within a session. A message is delimited by an End Of Message (EOM) event that can be selected from several supported by the PCU. When the sending PCU has detected a local End of Message condition, the current buffer is closed and transmitted. A buffer can contain both data and an EOM signal, but the EOM signal is the last that can be placed in the buffer. The EOM signal takes precedence over the two other conditions: buffer full and expiration of the idle timer. Several buffers or packets may constitute a logical message. Message length can be specified by the device and is not limited by buffer size.

The user can specify one of a series of EOM events, or a message length, to be recognised:

• *none*. If *none* is the EOM parameter selected, an End Of Message event is not sensed.
• *newline*. Each newline character received from the device signals an EOM. The newline character can be defined by the device. When the device transmits the newline character to the PCU, the PCU places the newline character in the current buffer, closes the buffer and transmits it as a packet labelled EOM.
• *control character*. When the device transmits any ASCII control character, the PCU places that character in the current buffer, closes the buffer and transmits it as an EOM packet. This condition ensures prompt delivery of ASCII control characters.
• *message length*. When the count of characters received from the device reaches a specified number, the internal buffer is closed and transmitted as an EOM packet.

NEWLINE EXPANSION
The device can request the local expansion of input newline characters in order to provide for automatic line feed. If newline expansion mode is enabled, when a newline character is received from the device the specified expansion will be echoed to the attached device regardless of the setting of the echo command. Only the newline character is sent to the remote PCU.

The device can specify whether this feature is enabled, and what character is used for the expansion.

COMMAND MODE

The PCU enters command mode after the user device issues a command delimiter. The delimiter can take one of the following forms:

- A string of two ASCII characters.
- The *break* signal.

The user must specify which form to use, using the *command* command. The default is the *escape* character (hexadecimal 1B), followed by the *delete* character (hexadecimal 7F). To determine what a PCU port's command entry sequence is, type the *status* command and note the hexadecimal values which appear next to the *command* listing.

The PCU normally echoes command characters locally and supplies prompts and status information. The device can disable this output, if desired. When command echo is disabled by issuing the *quiet* command with the parameter *on*, no command characters or command prompts are issued by the PCU, except for information returned in response to *help* and *status* commands. The *quiet* command is most useful for interfacing PCUs to computer communication ports so that PCU-issued prompts and messages are not received by the host, and so the host software does not have to be modified to interpret the prompts and messages.

Ports that have *quiet* enabled may appear inoperative. For this reason, the *help* and *status* commands always cause a display, regardless of the *quiet* mode setting.

BAUD RATE

The device can select a baud rate from the range between 19.2 kbps and 75 bps. As a result of this command, the PCU port begins sending data to the device at the selected baud rate, and expects data to arrive at that rate from the device.

PARITY RATE

The attached device uses the *parity* command to specify the parity applied to characters output to the device. If parity is *none*, 8-bit data is passed to the device as it was received. If parity is *odd* or *even*, parity is generated for 7-bit data. Incoming parity is never checked by the PCU.

STOP BITS

The device uses the *stops* command to specify the number of stop bits used on the EIA interface. The number can be 1, 1.5 or 2 bits.

TIMEOUT

Using the *timeout* command with a parameter between 1 and 30 min disconnects the port if no input or output to the device has occurred during the specified interval. Entering a parameter of 0 disables the automatic disconnect feature.

PUNIT

The *punit* command identifies the unit, and optionally the port, to which a permanent session is to be made. Parameters are the same as those for *call*. Using *unitld* and *portld* provides a fully qualified address; the *unitld* alone provides a rotary address.

PCALL

The *pcall* command determines if permanent sessions are to be established, and under what conditions. Calls are made to the unit and port specified by the *punit* command. Depending on the mode selected, sessions can be established automatically or when prompted. *Pcall* can be set to *on*, *off* or *auto*.

Table 14.2 Port specific data handling attributes.

Attribute name	Possible values	Default
Local flow control mode	none, EIA, xon/off	xon/off
Xon character	Any ASCII	DC1
Xoff character	Any ASCII	DC3
Character idle timer	10 to 2550 ms	50 ms
Command mode delimiter	none, break, sequence	sequence
Command mode characters	Any ASCII	ESC,DEL(1B, 7F)
Command mode response disable	on or off	off
Baud rate	75, …, 19200	19200
Parity	none, odd, even	none
Stop bits	1, 1.5, 2	1
Message length	0 to 255	0
Newline character	Any ASCII	ASCII CR
Newline expansion character	Any ASCII, or none	none
Local echo enable	on or off	off
Message delimiter	none, newline, control, length	none
Autobaud (pin enable)	on or off	off
Pcall	off, on, auto	off
Timeout	0–30 min	0 (disabled)
Punit	any unit and port, or rotary	0000

LOCAL ECHO

The *echo* command controls whether the PCU port echoes to the attached device the data presented to the PCU for transfer to a remote device. Echo can be set to *on* or *off*.

AUTOBAUD PIN ENABLE

The *autobaud* command enables and disables the functioning of the autobaud (AB) pin on the port's RS-232 connector. Autobaud can be set to *on* or *off*.

PORT-SPECIFIC DATA HANDLING ATTRIBUTES

Table 14.2 summarises the PCU attributes specific to each port and all sessions attached to the port.

PORT SPECIFIC DATA HANDLING COMMANDS

The following commands control the data handling attributes of each port on a PCU:

autobaud. Enables or disables the autobaud (AB) pin on the port's RS-232 connector. A double reset using the button on the PCU's rear panel will still cause all ports of the PCU to autobaud, regardless of the setting of this command.

baud. Specifies the baud rate to be used on the port.

command. Specifies the command mode entry string.

echo. Enables/disables local character echoing only when the PCU is in data transfer mode. The *quiet* command controls echo in command mode.

eom. Specifies the type of end-of-message processing to be done by the session.

expand. Both enables/disables newline expansion and specifies the character to be used for expansion.

flow. Specifies the method of local flow control to be used at the port.

idle. Specifies the value of the intercharacter idle timer.

newline. Specifies the character to be used for newline recognition.

parity. Specifies the parity to be added to each byte delivered to the user device.

pcall. Specifies whether a permanent session is to be established automatically, or in response to an external event.

punit. Specifies the unit and port or rotary group to which a permanent session is to be established.

stops. Specifies the number of stop bits to be added to each byte delivered to the user device.

timeout. Provides for automatic session termination if no data has been sent or received within a specified interval.

xoff. Specifies the character to be used to stop data transmission to (and from) the user device.

xon. Specifies the character to be used to start data transmission to (and from) the user device.

Signalling

The PCU can generate an out-of-band signal to a remote PCU and device. This signal is termed an *interrupt* control signal and is communicated across the session. Data packets buffered either in the source or destination PCU are not affected and the interrupt signal is not delivered in sequence with any buffered data. The signal is delivered to the destination port as a *break* – a 250 ms space on the RS-232C Receive Data (RD) pin.

The *interrupt* can be generated either by explicit command at the source PCU port or by the input of *break* on the source RS-232C TD pin. The latter is enabled if the command mode entry string has not been specified as *break*.

Diagnostic and maintenance services

Each PCU contains a set of software self-test routines for first level diagnosis of hardware problems. These diagnostics:

• Verify the condition of major LSI components.
• Checksum the ROM memory.
• Checksum the RAM memory.
• Test the buffer RAM memory.
• Echo test the modem.

LocalNet 20 PCU construction

EXTERNAL DESCRIPTION
The LocalNet 20 PCU is packaged in a stand-alone cabinet containing all required digital, analogue, interface and power electronics to connect two (20/100) or as many as eight (20/200) asynchronous terminals or computer ports to the LocalNet. Physically resembling a conventional point-to-point modem, the PCU is a small enough unit to sit near the terminal which it attaches to the network.

The front panel of the 20/100 PCU contains two multi-colour light emitting diodes (LEDs) used to display summary status information concerning the PCU and its activities. The front panel of a 20/200 PCU contains an additional eight LEDs, one for each user device port. A port LED is green when a session exists on that port, and unlit when no session exists. The port LED is red if no session exists and if the port will not accept an incoming call for one of the following three reasons:

- *DTR* control is *on* and the DTR signal is not asserted.
- *Listen* is *off*.
- *Maxsession* is 0.

The rear panel of the PCU contains connectors for:

- Power.
- DB-25 connectors for user devices (two on a 20/100).
- reset momentary pushbutton.
- A type F coax CATV fitting for attachment to the broadband cable system.

In addition, the rear panel of the 20/200 contains eight DB-25

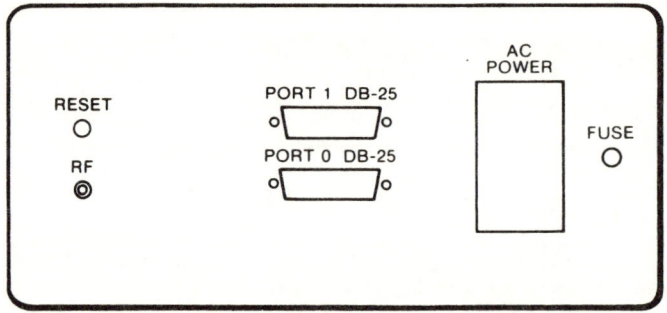

Fig. 14.6 LocalNet 20/100 PCU rear panel.

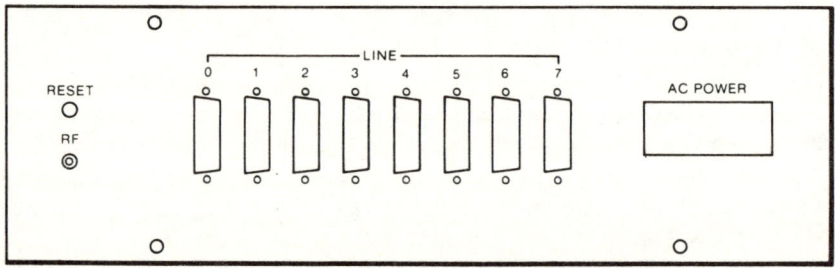

Fig. 14.7 LocalNet 20/200 PCU rear panel.

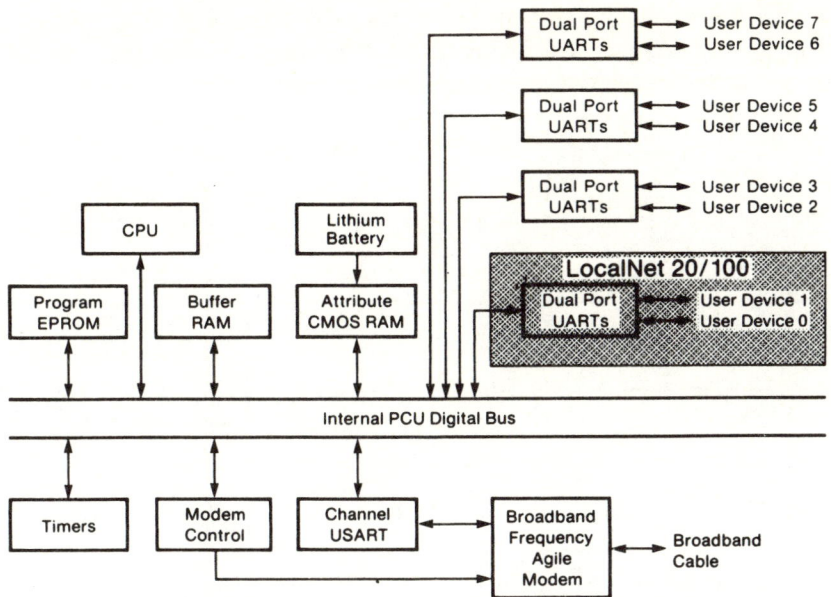

Fig. 14.8 LocalNet 20 PCU hardware block diagram.

connectors for devices. The rear panels are shown in Figs. 14.6 and 14.7.

The *lower* DB-25 connector for a 20/100 is user device port 0 and the upper is user device port 1.

Figure 14.7 represents the rear panel of a 20/200 PCU.

The *left-most* connector on a 20/200 rear panel is user device port 0; the connector at the far right is user device port 7.

HARDWARE DESCRIPTION
Internally, a LocalNet PCU is constructed from two printed circuitboards: a digital processing board and the analogue broadband modem board. An overall functional block diagram is given in Fig. 14.8. The PCUs contain static and dynamic RAM, and use static CMOS RAM, with a battery back-up, for storage of internal configuration and PCU and port attribute information.

INTERNAL SOFTWARE ARCHITECTURE
The LocalNet PCU software is constructed as a set of cooperating sequential processes, scheduled by a central event processor in response to communication events. In general, the software is organised to reflect the LocalNet protocol architecture, with two processes for each protocol layer. As outlined in Fig. 14.9, data

bytes (either received from the network or from user devices) pass through a sequence of protocol interpreting processes before delivery to the user device or the network, respectively.

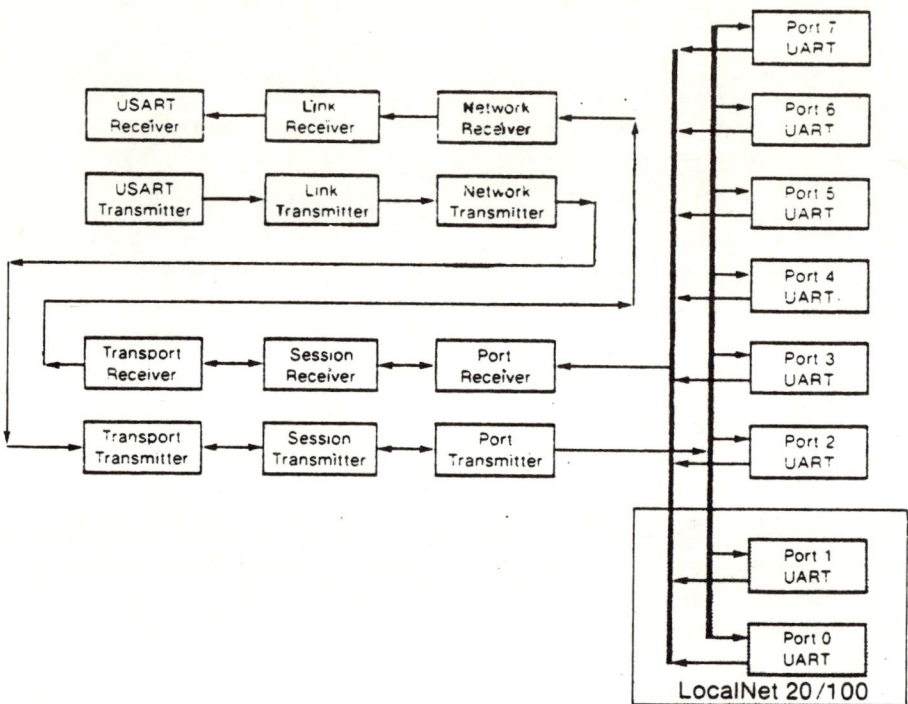

Fig. 14.9 LocalNet 20 PCU software architecture.

Elements	Functions
Tveter	Central retransmission unit
Tbox	Terminal interface
Tmux	Computer and/or terminal interface
Tbridge	Channel interconnection
Tlink	Interconnect LocalNets
Tgate	Connect to other networks
Network	Monitoring
Control	Diagnostics
Centre	Reconfiguration
	Resource management
	Security

Fig. 14.10 Summary of LocalNet System 20.

CHAPTER 15

Packet Communication Protocols

Devices connected onto a packet communication network must have some intelligence, usually provided by a microprocessor system. A number of procedures must be defined, both in the communication subnet and in the computing systems around it, to ensure the reliable execution of user tasks. These procedures entail the exchange of information between geographically separated hardware, and are implemented through software processes. Two such processes, which are usually separated physically, co-operate to implement the procedure.

Two processes running in different geographical locations must use the exchange of messages to co-ordinate their action and achieve 'synchronisation'. This message exchange must follow carefully designed procedures, called *protocols*. Their main characteristic is the ability to work when the timing and sequence of events can be unknown and where transmission errors are expected.

Protocol functions are accomplished by the exchange of messages between processes. The format and meaning of these messages form the logical definition of the protocols – *syntax*. Rules of procedure determine the actions of the processes co-operating in the protocol – *semantics*.

The protocol structure or architecture is hierarchical, consisting of a number of layers. The modularity of this approach minimises the interdependency between the various components of the network. The definition of the protocol structure makes no assumptions as to how the functions are implemented or distributed on network components.

This layered structure is commonly referred to as an *onion skin* architecture. Figure 15.1 is a schematic diagram of this approach. In this diagram layer n provides a service to layer $n + 1$, its 'user'. To do this it utilises the functions of the $n - 1$ layer to communicate with other entities at layer n in the network. Protocols are the communication rules between entities at the same level. The structure of the levels above or below n is not known to layer n itself. Each layer implements a set of functions to offer an enhanced service to the layer above it.

Across the interface between the n and $n-1$ layers control information and data are exchanged. Control information consists of messages between the two layers sent to ensure the correct operation of the interface, to request a specific task to be executed by the lower level and to report on progress or completion of a task to the higher level. The data passing across the interface are either control messages associated with the protocol and destined for the remote n level process, or data to be delivered to the remote $n-1$ level. A process therefore has three interfaces: two physical ones between the higher and lower levels, and a logical one with its partner in the protocol.

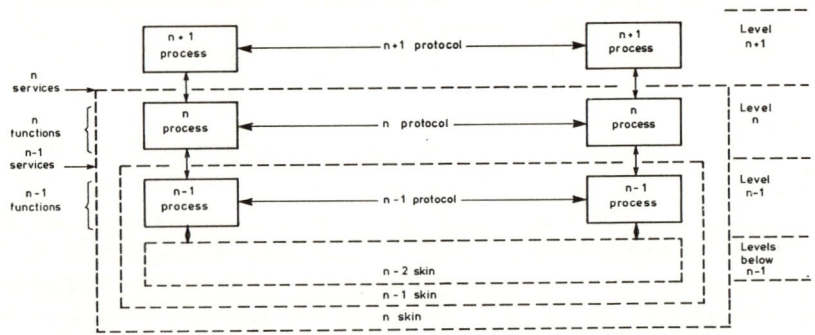

Fig. 15.1 Hierarchical protocol architecture.

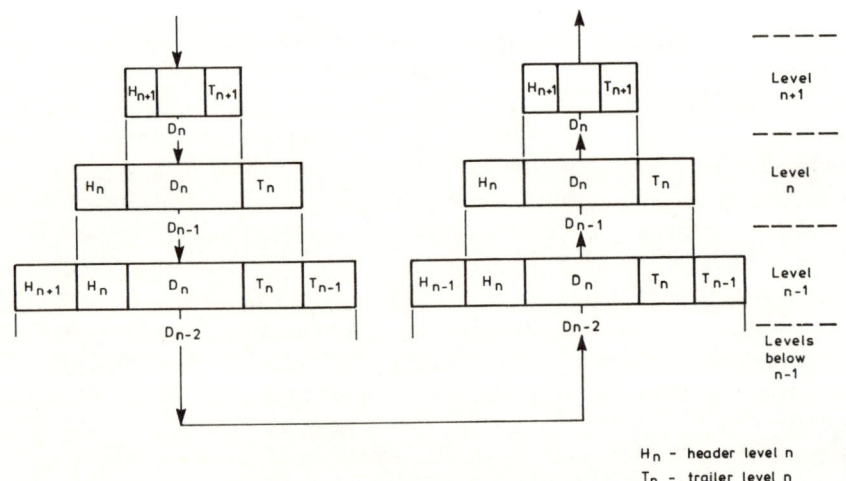

H_n – header level n
T_n – trailer level n

Fig. 15.2 Message structure.

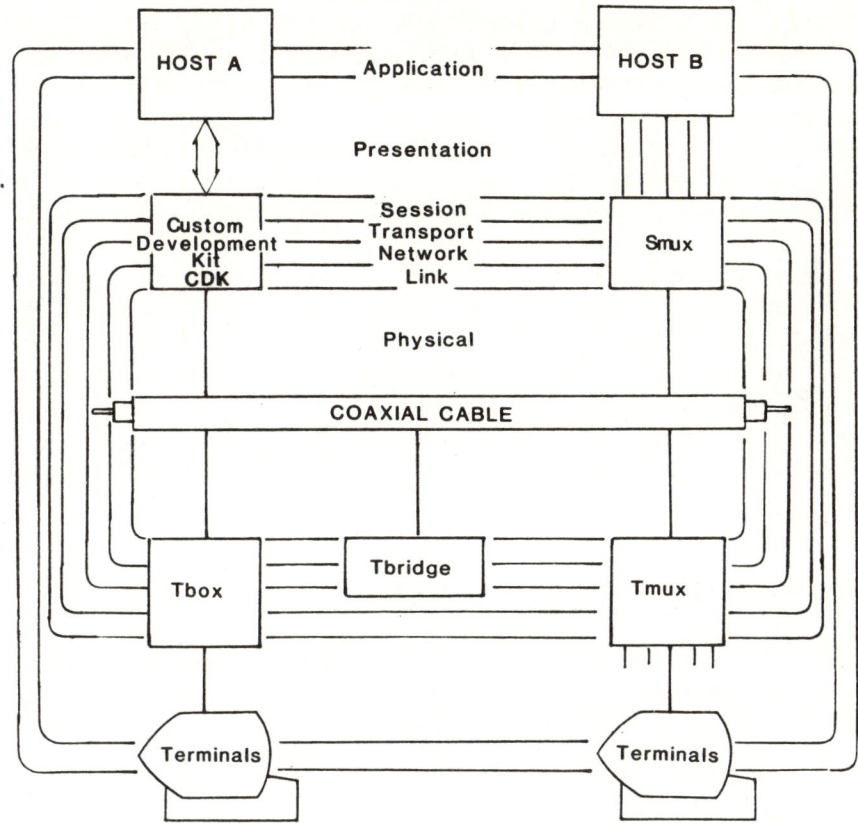

Fig. 15.3 LocalNet network architecture.

Whatever the functions or contents of messages sent from the higher to the lower level for delivery to the remote process, these are transparent to the lower layers.

The onion skin architecture leads to a layered structure of messages within the network as shown in Fig. 15.2. The header added at each level is associated with the operation of that level of protocol and can, for example, include numbering for sequence and error control and source identification. As the message progresses through the lower levels, more information is added. When the packets rise through the levels at the remote end, the headers and trailer information are stripped, processed and the data passed to the next level.

The principles of layered network protocol architectures are now well established and it would be instructive to study an overview of a broadband local area network protocol architecture. One example of a broadband local area network protocol architecture is incorporated in the Sytek Inc LocalNet product family (see Fig. 15.3). This architecture supports communication between user devices attached to one or more broadband local networks. The protocols are primarily intended to support communication on a broadband cable system, and between such systems; however, the architecture in fact defines a general communications system applicable within a variety of network environments.

Using the terminology of the ISO Reference Model[1], LocalNet provides communication services up to the presentation level. The architecture includes a 'connection oriented' end-to-end transport level protocol on top of a 'datagram oriented' network level protcol; this basic design is shared with other well known architectures such as TCP/IP[2] and XNS[3]. Here we will describe how the multi-channel broadband environment affected the specific designs of the LocalNet protocols.

THE LOCALNET ENVIRONMENT

A LocalNet network consists of a family of broadband packet switching local network products. The primary members of the product family are *packet communication units* (PCUs), which provide user devices with access to the cable network. Each PCU includes a frequency agile RF modem which allows transmission and reception of digital signals over a frequency derived channel on the broadband cable system. User devices physically attach to the network via access ports on the PCUs. These ports take a variety of forms depending on the type of PCU and type of user device. For example, these access ports can be RS 232C interfaces or a parallel DMA interface into host computers. In the latter case, the ports via which the user device accesses the Local Net services are logical ports multiplexed across the parallel interface.

Other members of the LocalNet product family do not directly attach to user devices, but are used to enhance the total network service. Of particular importance here is the interchannel 'bridge' and the intercable 'link' units. These devices extend the connectivity of a LocalNet so as to allow communication among all devices attached to all channels, and among devices which may even reside on distinct cable systems separated by long distances. The bridge device essentially provides a packet switch between channels, while the link device plays an analogous role between cable systems.

Figure 15.4 illustrates a generalised LocalNet network structure comprising two separate cable systems. The individual broadband channels of each cable system are joined by bridges and the two cable systems are joined by a link.

THE LAYERED LOCALNET PROTOCOL ARCHITECTURE

A full layered protocol architecture supporting user data communications is implemented in the digital logic and software within a PCU. This hierarchical LocalNet protocol architecture is layered in a manner similar to the ISO Reference Model for Open Systems Interconnection[1]. Table 15.1 depicts the ISO layers with the corresponding LocalNet protocols.

At the lowest level of the LocalNet architecture, a set of *physical* transmission media reside. In the case of LocalNet PCUs, the physical layer is realised as a set of 128 Kbps broadband channels. Other physical media such as high speed fibre optic cables may be used to connect separate cable systems into one logical network via the link product described above, and thus these media may also form part of the LocalNet physical layer.

The next higher layer in the LocalNet architecture consists of the

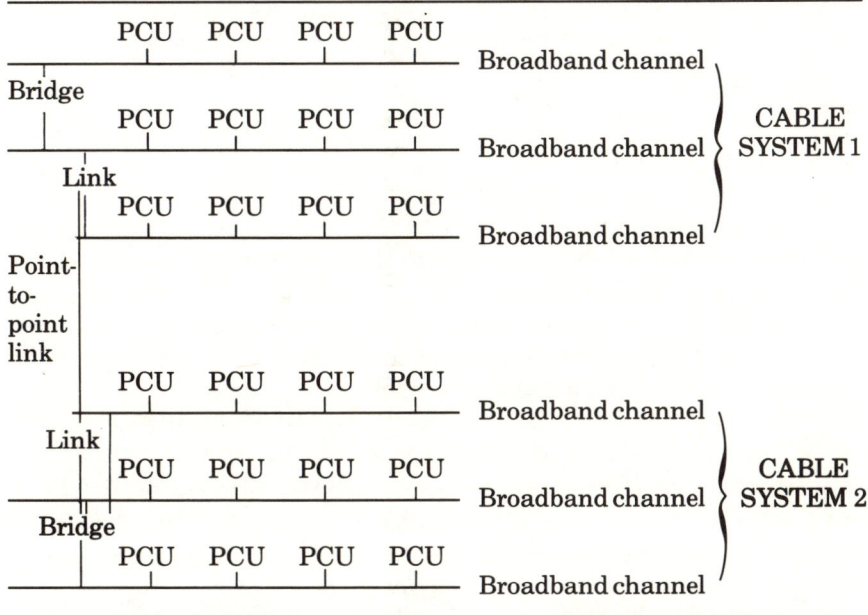

Fig. 15.4 Generalised architecture of a LocalNet network.

Table 15.1 ISO layers and the LocalNet protocols.

ISO	LocalNet
Application	(User-specific applications)
Presentation	Datagram, stream transfer, others
Session	Session management protocol (SMP)
Transport	Reliable stream protocol (RSP)
Network	Packet transfer protocol (PTP)
Link	Link access protocol (LAP)
Physical	Broadband channels, point-to-point

link protocols. These protocols are responsible for managing a device's access to the bandwidth of a given physical medium, appropriately framing the device's transmissions and detecting transmission errors. Link protocols within the LocalNet architecture need not correct or recover from detected bit errors, since higher level protocols will perform end-to-end retransmission of packets when necessary. Consequently, the LocalNet link protocols are responsible for detecting bit errors (e.g. via a frame check sequence technique), and may simply drop any packet received with errors.

The primary link level protocol within the LocalNet architecture is called the *link access protocol* (LAP). This protocol manages the link layer functions for the broadband channels. LAP's access mechanism is CSMA/CD (carrier sense multiple access with collision detection), and it uses standard HDLC framing and error detection techniques. Other link level protocols are required to manage the transfers across a point-to-point (i.e. non-contention) link between cable systems; for example, HDLC could be used across a fibre optic link. However, since the LocalNet protocol architecture requires only error detection (rather than full error recovery) at the link level, full HDLC (with link level retransmissions) is not necessary for the point-to-point links. Error recovery within LocalNet, as we shall see, is the responsibility of the transport layer.

The *network* level protocol within the LocalNet architecture is the *packet transfer protocol* (PTP). As with any network layer protocol within the ISO Model, PTP is responsible for routing. Within a LocalNet network, routing issues primarily concern the routing of packets between broadband channels within a single cable system, and routing of packets across point-to-point links between separate cable systems. As described earlier, this routing function is the responsibility of the bridge and link family of products. These devices are thus essentially PTP machines.

Using information maintained within internal routing tables, the PTP interpreters within bridges and links determine the next channel or point-to-point link across which a received packet should be forwarded. Of particular importance within LocalNet is the manner in which PTP accomplishes the initialisation and updating of these routine table entries. This is accomplished via a 'broadcast discovery' technique which takes advantage of the broadcast nature of the underlying broadband channels to accomplish this function in a distributed fashion with no need for centralised control.

PTP provides a datagram service, in which each packet is transferred across the network independently, with no delivery guarantees provided by mechanisms within PTP. Delivery guarantees are the responsibility of the *transport* layer within LocalNet, in which resides the *reliable stream protocol* (RSP). RSP provides virtual connections between PCUs. Each virtual connection is reliably sequenced and separately flow controlled, ensuring that a sender will not unnecessarily flood a receiver with data at a rate beyond the receiver's processing capacity. These reliable, flow-controlled connections provide the basic transport service used for applications such as file transfers and interactive terminal access. Much of the cost effectiveness of a local network can be traced to the sharing of a single transmission medium by multiple virtual connections; without a local network one typically requires many physical links to support these connection oriented services.

To provide this virtual connection service, RSP builds upon the datagram service of PTP, ensuring reliability via acknowledgements and retransmissions. The flow control function is handled via a 'window' scheme[4] in which a receiving RSP entity can signal to the sending RSP entity the current amount of data that can be accepted.

Each PCU can support multiple simultaneous RSP virtual connections, and in fact there can be multiple simultaneous connections between the same pair of PCUs. Each connection is identified by a 'connection identifier' (one for each side of the connection). Users of RSP can request the transmission of data over a particular connection by referring to that connection's local connection identifier.

The *session* layer protocol within LocalNet is called the *session management protocol* (SMP). SMP provides an extended addressing service out to the port level, allowing user sessions between user devices attached to PCU ports. Each port can be configured according to a set of parameters, such as the maximum number of simultaneous sessions to be allowed on the given port. SMP provides additional support for the establishment of user sessions, including a 'rotary' service in which a collection of LocalNet PCUs can be configured into a rotary group, with user call requests automatically directed to a free port within the rotary group. A 'privilege' concept also exists within SMP for certain types of PCUs. This concept allows PCUs designated as privileged to override the disablement of certain functions and commands on both local and remote PCUs. Finally, the LocalNet session layer is the location of other enhancements of the call establishment functions, such as encryption key distribution.

LocalNet services are exported to the users via a variety of *presentation* protocols. These protocols allow dissimilar devices to take advantage of the basic session service provided by SMP. For example, presentation level services are defined to support asynchronous ASCII terminals and bisync devices, and to provide block-mode DMA parallel interfaces for direct host computer attachment.

ADDRESSING WITHIN THE LOCALNET ARCHITECTURE

A LocalNet node is assigned a 16-bit identifier, the unitID, by the network's manager. Such a unitID must uniquely identify the node within the entire LocalNet installation, and will typically not change.

When a node is assigned to a specific channel (i.e. its modem is tuned to the appropriate frequency), the node must be assigned a link level address by the network manager. This 8-bit address, called a *LAP address*, is used by the LocalNet link access protocol (LAP) to identify uniquely all nodes on a given channel. LAP addresses are only unique within the scope of an individual channel, and it is entirely possible that two nodes on different channels may have the same LAP address (though their unitIDs will of course be different). The small 8-bit LAP address is used by hardware address recognition circuitry to filter out frames received on the LocalNet broadcast channels.

PTP is responsible for determining the proper LAP address to use if a particular unitID is the intended target of a transmitted packet. If the intended target is on another channel, the appropriate LAP address will be the LAP address of a bridge on the current channel which can then forward the packet on to the appropriate next channel.

Ports are the final piece of the LocalNet addressing scheme. As described above, ports are an SMP construct, allowing the RSP virtual connection service to be extended, so the connections are between user devices attached to ports rather than just between PCUs. Thus the address of a user device within a LocalNet network is given by the pair 'unitID, portID'. Note that a LAP address is not necessary to identify uniquely a device within a LocalNet installation.

OVERVIEW OF SPECIFIC PROTOCOLS

Working from the bottom up, we can now describe in overview fashion the specific functions and mechanisms within each of the major protocols within the LocalNet architecture.

Physical layer

The primary medium at the LocalNet physical layer is broadband cable.

Link access protocol

The protocol used at the link level on LocalNet broadband channels is the link access protocol (LAP). LAP provides a frame transfer service between any two nodes which are tuned to the same channel.

Nodes are identified at the link level by the 8-bit LAP address. As discussed earlier, LAP addresses do not uniquely identify nodes over the scope of an entire multi-channel LocalNet; this is the responsibility of the unitID address at the PTP level.

LAP provides basic framing (with data transparency) and error detection. Framing is handled by the same technique used within HDLC, in which frames are delimited by the bit pattern 01111110 (a 'flag'), and data transparency is achieved by automatically inserting a zero after a sequence of five ones (and deleting such a zero upon reception). Error detection is accomplished through use of the CCITT polynomial cyclic redundancy check mechanism (CRC-16). LAP does not attempt to recover from detected bit errors – this is the responsibility of the higher level protocols.

Access to the shared channel is managed via carrier sense multiple access with collision detection (CSMA/CD), similar to that used in the Ethernet network[5]. Each node which has a frame ready for transmission will first listen to the channel to check if any other node is currently transmitting. If the channel is free, the node may begin to transmit on his 'send' frequency. The cable system headend will convert these transmitted signals to the 'receive' frequency. While transmitting, the sending node continues to listen to the receive frequency, checking to see if the initial received bytes of the packet are the same as the initial transmitted bytes. If the initial transmitted and received bytes match, then it is inferred that no collision has occurred, the sending node stops listening to the network, and the remainder of the frame is transmitted. If they do not match, then it is inferred that another node is also transmitting, causing a collision; in this case, the frame's transmission is aborted and it is scheduled for retransmission.

Retransmission of a frame which experiences a collision will be attempted after a binary exponential random timeout period[5]. The timeout period for a retransmission attempt is an integral number of 'slot times', where a slot time is essentially the maximum end-to-end propagation delay on the cable system. Note that since broadband signals must traverse the cable twice (up to the headend and then back), the maximum end-to-end propagation delay occurs between two stations which are furthest from the headend. (For a system with a longest cable leg of 30 km, this yields a slot time of 300 µs). The retransmission of a collided frame will be delayed for n slot times, with n chosen randomly between 0 and 2^k where k is the number of transmissions attempted thus far for this particular frame.

A frame (shown in Fig. 15.5) consists of a delimiting flag character, a local destination link level address, a data field, and a CRC field. The transmission of a frame is preceded by the transmission of N opening flag characters, where N depends on the network characteristics. These flags serve the purpose of establishing byte synchronisation for the correct delimiting of incoming bit sequences.

The end of a frame is signified by a closing flag character. Data transparency within the frame is ensured by automatic zero bit insertion and deletion procedure as specified in the HDLC specifications on which this frame format is based.

The LAP address field (8 bits long) contains the destination link level address of the node on the channel which should receive this frame. Note that this may or may not be the ultimate destination for the PTP packet embedded within the frame. A value of 11111111 for the LAP address indicates 'broadcast'; frames which include such an

Opening flags	LAP address	Data	CRC – 16	Closing flags
8 bits each	8 bits	variable	16 bits	8 bits each

Fig. 15.5 LAP frame format.

address will be received by all nodes (a feature which is used to advantage by the PTP 'discovery' mechanism).

The data field contains the packet passed down from the higher level protocols.

The CRC field is 16 bits long and is produced via a hardware CCITT polynomial generator. The CRC is generated for all bytes within the frame excluding flag characters (and inserted zero bits for data transparency). A 16-bit CRC is adequate due to the inherently low bit error rate of the broadband transmission medium (typically 1 error in 10^9 bits).

At least one closing flag is sent in order to terminate properly the HDLC-like frame. A small number of additional flags may be sent depending upon the hardware characteristics of the PCU.

Packet transfer protocol

PTP provides the inter-channel and inter-cable routing functions within the LocalNet architecture. PTP is a datagram protocol, essentially treating each packet independently and providing no reliability or flow control mechanisms.

BASIC PACKET TRANSFER

A packet will be forwarded from one PTP interpreter so another until its final destination is reached. When the source and destination are on the same channel, no intermediate PTP interpreters will be involved in the transfer. However, if source and destination nodes reside on different channels or cable systems, then intermediate PTP interpreters (within bridges and/or links) are required for packet forwarding.

The PTP implementation within a bridge maintains a 'location table' which associates unitIDs with channel numbers and LAP addresses. Packets are forwarded between channels based on the information stored within the location table.

If source and destination nodes reside on the same channel, the packet will be embedded within a LAP frame that includes the destination's LAP address. The destination's LAP interpreter then picks the frame off the channel and passes it up to the destination's PTP interpreter. Typically the packet will then be passed up to the appropriate transport entity within the destination node as specified in the packet's PTP header.

However, if the source and destination nodes are not on the same channel, the source LAP interpreter must embed the packet within a LAP frame that includes the LAP address of the appropriate intermediate bridge. That bridge's LAP interpreter will pick the frame off the channel and pass it up to the bridge's PTP interpreter. Now the bridge's PTP can consult its location table to determine the appropriate channel and LAP address for the destination unitID given in the packet header. The packet is then passed to the appropriate channels's LAP interpreter and embedded within a LAP frame with the next LAP address obtained from the bridge's location table. Note that this new LAP address would be the LAP address of the ultimate destination if the destination resided on that channel, but would be the LAP address of another bridge if the route required multiple 'hops'.

BROADCAST DISCOVERY

The previous section described how packets are transferred by PTP interpreters when the appropriate route is already known. This section gives an overview of the 'broadcast discovery' mechanism which determines packet routes.

Initially, no node knows the location of any other node; here 'location' means a node's channel and LAP address. The location table entries in all bridges are initially empty.

When a transport entity such as an RSP interpreter wishes to communicate with another node, it passes the data to its PTP interpreter with the destination node's unitID. If the PTP does not know where the destination unitID can be found, it tells its local LAP interpreter to broadcast the packet. All nodes on the channel will pick up the broadcast and check the intended destination unitID. If one of them is the intended destination it will respond with a PTP packet that includes its LAP address. The initiating node's PTP can then address future packets directly to the destination node's specific LAP address, so no further broadcasts are required.

If the destination node is not on the same channel, however, an intermediate bridge will pick up the broadcast packet and must forward it via broadcasts on the other channels to which it is connected. These broadcasts will be picked up directly either by the intended destination node or by other intermediate bridges. Eventually the destination node is reached, and it responds over its channel with a packet containing its LAP address. Bridges along the route update their location tables to reflect their new knowledge of the destination's location, and future packets sent between source and destination will not require further broadcasts.

SWITCHING EFFICIENCY

In the 'steady state' situation, with a route between source and destination established and reflected in the various bridge location table entries, the bridges play the role of packet switches between channels. A key concern is the efficiency (in terms of processing overhead) of this packet switching operation.

Each time a bridge forwards a packet, it must determine the appropriate channel and LAP address for the packet's next hop, based on the location table entry for the destination unitID contained in the packet's PTP header. The efficiency of these table look-ups is a major factor determining the overall performance of the network. For example, in a large installation a bridge may need to have location table entries for several thousand nodes. If the table were a simple list which required a linear search, each packet switching operation could require an average of over 5 ms (based on instruction timings typical of current generation 16-bit CPUs). Since end-to-end propagation delays on a broadband cable system range are at most of the order of 500 μs, it is clear that the packet switching operation could be the major component of end-to-end delay. Clearly a more efficient approach is required for these table look-ups; of course, it is also desirable to minimise the size of the associated location table data structure due to memory resource limitations within the bridge.

PTP is designed to facilitate these table look-ups by using additional information contained in packets. Incoming packets in fact contain not just the destination unitID but also the actual index of that unitID's location table entry. Consequently, the bridge does not have to scan the entire location table on each packet switch operation. This is accomplished by allowing bridges to pass a 'logicalID' back to a source node, asking it to include this logicalID in a PTP control field for every packet going to a given destination. No interpretation of the logicalID is made by the source node; it is simply a value which is placed in outgoing packets to help the receiving bridge quickly locate the appropriate destination node's entry within its location table.

The values used by a bridge for logicalIDs are in fact direct indexes into its location table. This mechanism substantially improves bridge packet switching efficiency.

Destination unitID	Source unitID	Control	Data	Trailer (optional)
16 bits	16 bits	8 bits	variable	40 bits (if present)

Fig. 15.6 PTP packet format.

Destination logicalID	Source logicalID	Source LAP address
16 bits	16 bits	8 bits

Fig. 15.7 PTP trailer format.

PTP PACKET FORMAT

Given the above description of PTP functions, the format of the PTP packet (shown in Fig. 15.6) can now be explained. The trailer is optional and its presence is indicated by a bit in the control field. The format of the trailer is shown in Fig. 15.7. In Fig. 15.6 the destination and source unitIDs indicate the addresses of the ultimate destination of the packet and of the packet's originator. The control field includes several subfields which indicate the presence or absence of the optional trailer, and the higher level protocol to which the packet's data field belongs.

The optional trailer (Fig.15.7) is included on broadcast discovery packets and on inter-channel packets. (Note, however, that the last bridge in a PTP route strips off the trailer before delivering an inter-channel packet to the final destination PCU.) The trailer includes the logicalIDs which give the index into the bridges' location tables.

Reliable stream protocol

OVERVIEW OF RSP

RSP – the LocalNet reliable stream protocol – provides error free virtual connection services to its users. Assuming only an unacknowledged datagram service as its lower level protocol, RSP accomplishes its reliability through end-to-end acknowledgments and retransmissions.

RSP resides at the transport layer of the LocalNet protocol architecture, and uses the packet routing services of the network layer PTP protocol. RSP builds a reliable stream service on top of PTP's services by attaching sequence numbers to each packet and providing for their acknowledgement and (if required) their retransmission.

RSP's retransmission mechanisms have been designed under the assumption that the lower network service may occasionally lose packets, but will in general not misorder them. This will typically be true within LocalNet because no alternate routes are provided for packets belonging to a given connection. It is important to note that, while RSP will still function correctly with a network service that occasionally misorders packets, RSP's sequencing and retransmission mechanisms have been optimised for an assumed network environment where the misordering of packets is rare. This assumption proves valid in the broadcast environment of a broadband local area network. This assumption remains valid even when a large number of alternate routes exist between bridged channels; this is because RSP sessions will choose one route through the bridges at the time the session is established and thereafter use only that route, thereby avoiding packet duplication and resultant misordering.

RSP also includes mechanisms which ensure that a data source does not flood a receiver with more data than it can handle. This is accomplished through a sliding window scheme[4], whereby a receiver communicates to a sender its current ability to handle new packets. This function is commonly referred to as *flow control*.

These reliability and flow control services are made available to RSP's users through *connections*. Users of RSP can establish a connection with each other, exchange data over the connection, and terminate connections when no longer required. RSP connections are separately flow controlled by RSP, and data which is sent over a particular connection will be delivered to the other end in a properly sequenced, duplicate-free fashion.

Connections are initiated at the request of a session layer user through the transmission of an RSP 'openRequest' packet. Upon receipt of such an openRequest packet, an RSP entity will return an 'openAck' acknowledgement packet if it has the resources to handle the connection; otherwise it will return an 'open*N*ack' packet. Through the information exchanged with this handshake, the two sides agree on a means of identifying the new connection (via 'connection identifiers') and on various connection parameters. Once this connection establishment control packet exchange is complete, data can be exchanged between the users – the session layer entities on whose behalf the connection has been opened. This data exchange will be reliable and flow controlled.

RSP packages user data to an RSP packet, attaching RSP control information in a header. An identifying sequence number is placed in each RSP packet. The sequence number is incremented with each new packet transfer over a given connection which contains data. This enables the receiving RSP to determine the proper order in which the received packets were transmitted.

Every packet exchanged across an RSP connection also includes a window field within its header. An RSP entity fills this field with the number of packets which it is currently willing to receive. A sending

RSP entity must not send more packets than it knows the receiver will accept based on the most recently obtained value of the receiver's window.

RSP connections are ordinarily 'closed' via an exchange of control packets at the request of one (or both) of the connection's users. Connections may be terminated prematurely if one of the PCUs 'dies' while the connection is in progress. To ensure a timely recovery from such a half-open connection, an RSP interpreter will periodically transmit a 'keep alive' packet which informs its partner that it is still active. If no packet is received over a connection for a specified time, the remote side is presumed dead and the user is informed that the connection has been aborted.

RSP RETRANSMISSION STRATEGY

RSP differs from other transport protocols (such as the Department of Defense TCP[2], and the ECMA-72 Transport Protocol[6]) in its acknowledgement and retransmission strategy. These design differences can be traced to the intended environment – local networks – and the service provided by the LocalNet PTP network layer protocol:

(1) The underlying broadband transmission medium provides relatively low bit error rates. Consequently, RSP is optimised under the assumption that most packets will be correctly delivered.

(2) The LocalNet product family includes inexpensive packet communication units which must include an RSP implementation. Consequently, RSP must be capable of being implemented on simple microprocessors with limited memory.

(3) Except in rare circumstances, the routing service provided by PTP does not misorder packets. Packets may be lost, but they will typically be delivered to a receiver in correct sequence, with occasional gaps due to lost packets. Consequently, RSP can be optimised to take advantage of this lower level service.

These considerations have yielded a simple design for RSP in which packet retransmissions are initiated primarily via a combination of a 'negative acknowledgement' scheme and a 'positive acknowledgement with timers' scheme.

When a packet is transmitted, it is placed on a 'retransmit queue' by the sender, where it stays until its receipt has been acknowledged by the remote RSP. A receiving RSP entity includes positive acknowledgements of received packets within the headers of packets going in the reverse direction. The sender of a packet can request that the receiver return an acknowledgement by the inclusion of a 'poll' bit within a packet header. (Typically this occurs if the user has indicated that this is to be the last packet within a sequence of logically related packets, or if the transmitted packet is at the boundary of the current flow control window.) When such a poll-containing packet is transmitted, a 'retransmit timer' is started; upon timer expiration a packet retransmission procedure is invoked. By starting timers only on poll transmission (rather than on every packet transmission), the RSP implementation is simplified, satisfying (2) above. The inclusion of such a poll mechanism distinguishes RSP from both TCP and the ECMA protocol, and can be traced to a desire to simplify implementations – in particular to simplify the rules followed by a receiver for generating acknowledgements.

RSP's negative acknowledgement scheme is, however, the primary method by which packet retransmission is initiated. A receiving RSP entity will only accept data within a packet if the packet is the next one expected. This rule considerably simplifies the design of RSP implementations, satisfying (2) above. Due to (3) above, the loss of a packet can usually be detected via the receipt of another whose sequence number is not the next expected. If a packet is received which is not the next packet expected (indicating that the expected packet got lost), the RSP entity will generate a special control packet (a 'nack', or negative acknowledgement packet), which requests that the sender retransmit all unacknowledged packets.

If a sender transmits five packets, and the second one gets lost, it is not desirable for the receiver to send a nack upon reception of the third packet, another nack upon reception of the fourth, and a third nack upon reception of the fifth. If this were to happen, the sender would receive three 'commands' to retransmit packets 2 to 5, when in fact only one of these nacks should be acted upon. To prevent this problem, a receiving RSP entity follows a set of 'nack suppression rules', which together ensure that only one nack is generated in a situation such as the above. A 'transmission sequence' is defined to be a set of received packets whose sequence numbers increase. Thus, whenever packets are being retransmitted, a new transmission sequence is initiated. The nack suppression rules followed by a receiver ensure that only one nack per transmission sequence is generated.

The final aspect of RSP's retransmission strategy concerns the action taken upon expiration of the retransmit timer. Two approaches are possible:

(1) Retransmit all unacknowledged packets.
(2) Retransmit only the last unacknowledged packet.

Because of the presence of a negative acknowledgement mechanism, RSP takes the second approach. This distinguishes the RSP implementation from most TCP and ECMA-72 implementations. Taking into consideration the excellent error characteristics of the underlying medium and the 'non-reordering' packet delivery service of PTP, it is easily seen that retransmit timer expiration will be typically caused by loss of either the last transmitted packet or its acknowledgement. If an earlier packet was lost, the generation of a nack by the receiver will usually cause retransmission of all unacknowledged packets prior to expiration of the retransmit timer.

RSP thus provides standard transport connection services via mechanisms which have been optimised for the LocalNet environment. This has allowed implementation of typically complex transport layer functions within units based on simple microprocessors.

RSP PACKET FORMATS

An RSP packet consists of a number of fixed fields, called the *header*, followed by data bytes. The first byte of the header is called the *send control field*. the packet type contained in the send control field determines the format and meaning of the other header fields. The major packet types include:

- Open request (openRequest).
- Open acknowledgement (openAck).
- Open reject (openNack).
- Data.
- Retransmit request ('nack').
- Close.
- Closing.
- Closed.
- Interrupt.

The basic RSP packet format is illustrated in Fig. 15.8.

The send control field and the receive control field are present in all RSP packets. As noted earlier, the send control field contains a packet type indicator which determines the format and meaning of the remaining portion of the header.

The primary function of the receive control field is to communicate

Send control	Receive control	Destination connection identifier	Send sequence	Receive or user	Other control or user data
8 bits	8 bits	8 bits	8 bits	8 bits	variable

Fig. 15.8 Basic RSP packet format.

flow control information. This field includes a 4-bit sub-field which allows an RSP entity to indicate which packets it is currently willing to receive.

The destination connection identifier field is used to name the connection to which this packet belongs. The openRequest and openAck packets also include a source connection identifier field which must be exchanged by the two parties at the outset of a new connection.

The send sequence field contains the sequence number of the packet; the receive sequence field contains the sequence number of the next packet expected in the reverse direction (used as an acknowledgement).

The remaining fields are either additional control information specific to a given packet type (e.g. the source connection ID in an openRequest packet) or contain user data for an established connection.

Session management protocol

SMP – the LocalNet session management protocol – provides basic session management services for interactive users and computers attached to a LocalNet network. These services include user device addressing and support for the remote control of a device's interface parameters.

SMP constructs data communication sessions between communicating ports. These ports are provided by LocalNet PCUs as the access point for user devices. Both the characteristics of the session and the port can be adjusted for the particular requirements of the communicating user devices. The specific attributes of both sessions and ports are (possibly) unique for each user device and are not prescribed by SMP. However, SMP does provide a generic means for manipulating these attributes through invocation of PCU local commands.

The session service is constructed atop RSP transport level virtual connections that are in turn constructed from network level datagram packets. Each SMP session is constructed atop a single transport connection. Many sessions can be attached to each port.

SMP PARAMETER SETTING
A port is configured through the setting of certain parameters. SMP allows both the local and remote user to control these parameter settings. This is accomplished through commands to the PCU which can be separately enabled or disabled. All enabled commands are executable either by the local user device or by a remote device user to which a session is outstanding.

Typical sets of parameters and commands for a LocalNet asynchronous PCU are shown in Table 15.2. These parameters and commands are similar to those used in the CCITT X.3 and X.28 parametric terminal protocols[7]. The display of parameters is produced via a command called 'status'; the display of available commands is produced via a command called 'help'.

THE PRIVILEGE CONCEPT

A 'privilege' concept exists within SMP for certain types of LocalNet PCUs. This concept allows PCUs designated as privileged to override the disablement of certain functions and commands on both local and remote PCUs. Some examples of the effect of the privilege concept on the operation of LocalNet PCUs are:

- A privileged PCU can establish a session with an unprivileged PCU and execute any valid command, including disabled commands, on the remote PCU.
- A privileged PCU can override limitations on the number of sessions that can be established at a called or calling PCU port.

The privilege concept is commonly used in LocalNet for network administration and reconfiguration activities.

SMP ADDRESSING

In addition to the use of unitIDs, SMP portIDs are used to identify user device access points.

Table 15.2 Typical parameters and commands for a LocalNet PCU.

Parameters:

LocalNet 20/100 0C01 0000 1A V2

UNIT	0012,1	BAUD	4800	IDLE	5
GROUP	A	PARITY	NONE	EOM COUNT	0
LOCATION	08,01	STOPS	1	EOM CHARACTER	NONE
COMMAND	1B,7F	AUTOBAUD	OFF	NEWLINE	0D
LISTEN	ON	DCD CONTROL	OFF	EXPAND	NONE
PRIVILEGE	OFF	DTR CONTROL	OFF	XON	11
MAXSESSION	2	ECHO	OFF	XOFF	13
PCALL	OFF	QUIET	OFF	FLOW	XON
PUNIT	0000	TIMEOUT	0		

Sessions:
  (1) 0752,1★ ACTIVE
  (2) 1019,0 SUSPEND

Commands:

AUTOBAUD	BAUD	CALL	COMMAND	DCD	DISABLE
DTR	DONE	ECHO	ENABLE	EOM	EXPAND
FLOW	GROUP	HELP	IDLE	INTERRUPT	LISTEN
LOCATION	MAXSESSION	NEWLINE	PARITY	PCALL	PRIVILEGE
PUNIT	QUIET	REMOTE	STATUS	STOPS	SUSPEND
SWITCH	TIMEOUT	UNIT	XOFF	XON	

A LocalNet device address is a doublet: unitID, and portID. The unitID uniquely identifies the PCU desired, and the portID uniquely identifies both the physical user device attached to the PCU, and perhaps a logical access point within that user device. Many sessions can be associated with each access port and with each PCU. Each is uniquely labelled by a specific sessionNo.

LocalNet is commonly employed in so-called 'port contention' data switching applications, so two possible methods of opening a session exist. A request for a session to a PCU may specify only the unitID (partially qualified call) or both the unitID and a specific portID (fully qualified call).

On fully qualified calls, SMP attempts to establish a session to the specified port. If the port is idle and able to accept sessions, then the session is immediately established. If the port is busy or unable to accept sessions, then a 'port busy' message is returned to the user.

On partially qualified calls, SMP uses one of the rotary calling schemes described below. There are two available rotary calling schemes:

(1) *Sequential rotary.* In this scheme, the calling PCU makes a partially qualified call (session request) to another PCU. The called PCU will rotor through its ports, starting with port 0, until it either finds an idle port or it finds that none of its ports is idle. If none is idle, then the called PCU returns an 'all ports busy' message to the calling PCU. The calling PCU then increments the unitID by one and issues another partially qualified call. If the second call is to a PCU which has idle ports, then a session will be established to the first idle port. If the second call is to a PCU which is non-existent or to a PCU that cannot respond (e.g. power removed), then the calling PCU's response timer will expire and the rotary process will be terminated.

This scheme has the advantage of being simple and predictable. Its disadvantages are that it does not evenly distribute call traffic in the rotary (i.e. the PCUs with the lower unitIDs get the most calls), and that a non-respondent PCU (say unitID N) 'breaks up' the rotary by not allowing the calling PCU to access the remainder of the rotary (unitIDs $N + 1, + 2 \ldots$).

(2) *Random rotary with specified rotary size.* In order to address the disadvantages of the sequential rotary approach, a random rotary scheme was added to SMP. The basis of this scheme is that the number of PCUs in the rotary is specified in the session request command. The calling PCU then uses a randomised index along with the called unitID and the specified rotary size to select the beginning of the rotary group.

To provide flexibility, the sequential rotary scheme is available and is automatically invoked when the size of the rotary, X, is not specified.

SESSION INITIATION AND TERMINATION

A PCU responds to local and remote requests to initiate sessions. Local requests take the form of a user command.

A LocalNet PCU recognises both fully qualified and partially qualified (rotary) addresses. The PCU treats local session requests with partially qualified addresses as rotary session requests. The calling PCU attempts to locate an appropriate remote PCU by using one of the rotary schemes previously described. The called PCU responds to a request for a rotary session by constructing a session to the first idle PCU port that is enabled for incoming sessions or, if there are no free ports, it will return a message indicating 'all ports busy on this PCU' to the calling PCU.

SMP supports multiple sessions for each user device port. Also, a user device can disable all incoming session requests for its PCU ports.

SMP also supports termination of existing sessions. A local user device port may terminate a specified session. Sessions may also be terminated by the remote PCU.

DATA TRANSFER

Once a session has been established, SMP provides for the transfer of data between users. This transfer will be managed in a way dependent upon setting of the SMP parameters.

User data is packetised and transmitted by SMP at SMP's discretion. However, the user can control certain parameters which impact the packetisation. This includes SMP parameters relating to maximum packet size, packet-hold time, and end-of-message delimiter. For example, a PCU can be configured to interpret a certain user data character or character count as an 'end of message' indication. This indication will force the SMP to transmit a packet even if the packet is not full. For certain types of PCUs, the 'end of message' indication can also be passed across to the remote user. In this case, delivery of data contained in received packets will be forced upon detection of 'end of message'.

SMP sessions are constructed upon flow controlled RSP connections, and this internal RSP flow control is exported to the user according to the settings of certain SMP parameters. PCUs have the capability to flow control attached user devices via various software and hardware mechanisms. This allows user devices of differing transmission speeds to match speeds and provides reliable data delivery in the event of network or PCO congestion. For example, XON/XOFF can be used to flow control an ASCII/RS-232C interface, ensuring that the PCU is not flooded with user data when the network connection is flow controlled.

SMP provides a service to the user allowing the generation of breaks and interrupts to the remote user device. Upon receipt of an interrupt command for a currently active session, the SMP generates a transport level interrupt packet. Flow control does not affect the request, transmission or generation of interrupts at the destination PCU and user device. LocalNet RS-232C user devices are notified of an interrupt by the output of a 250 ms BREAK signal.

SMP PACKET FORMAT
The SMP packet formats fit the general format shown in Fig. 15.9.

tcontrol	Data/control
8 bits	Variable – up to 64 bytes for LocalNet PCUs

Fig. 15.9 General SMP packet format.

The *tcontrol* – type control – byte, shown in Fig. 15.10, is contained as the first byte of all SMP packets. It is used to identify the type of packet and carries as well certain control flags delimiting logical sequences of data packets. The flags include an 'end of message' (EOM) field which labels this SMP packet as an end of message packet. Depending upon the type of user device, this EOM indication may be exported to the destination user device port after all data in the packet have been delivered to the user device.

Flags	Type
2	6 bits

Fig. 15.10 tcontrol byte format.

The *type* field identifies the type of SMP packet. The currently defined packet types include:

- Data packet.
- Session request.
- Session accept.

- Session reject.
- Remote command
- Command response.
- Privileged remote command.

The data/control bytes in some SMP packet types may contain other type-dependent control information. For example, the session request packet includes source and destination portIDs within this field.

Presentation services

The LocalNet protocol architecture allows for a variety of presentation interfaces to user devices. These interfaces export the basic session services of SMP in a format and data presentation style appropriate for a particular class of device.

Among the presentation interfaces currently supported by LocalNet are:

- *Asynchronous ASCII RS-232C* appropriate for standard start-stop terminal devices and their associated host computer ports. This interface is provided by the standard LocalNet family of PCUs.
- *BSC 'Bisynch' interface* supporting the binary synchronous communications protocol between the LocalNet PCU and IBM or other bisynch terminals and controllers.

Other presentation interfaces can be added to this list.

In addition to the basic SMP services of session establishment and data transfer, each of these presentation interfaces provides the user with the capability of configuring the PCU port with device specific parameters controlling, for example, the type of parity used, the manner in which flow control is to be managed across the user device/PCU interface, or the conditions under which packets will be formed.

Summary of packet encapsulation

Now that all the protocol layers and associated packet formats have been discussed, it is appropriate to summarise the layer-by-layer packet formation, often referred to as 'encapsulation'. This process is shown in Fig. 15.11.

An examination of the various field lengths indicates that the maximum packet length is as follows:

85 bytes (680 bits)

The above counts only one opening and one closing flag, and assumes that the optional PTP trailer is present.

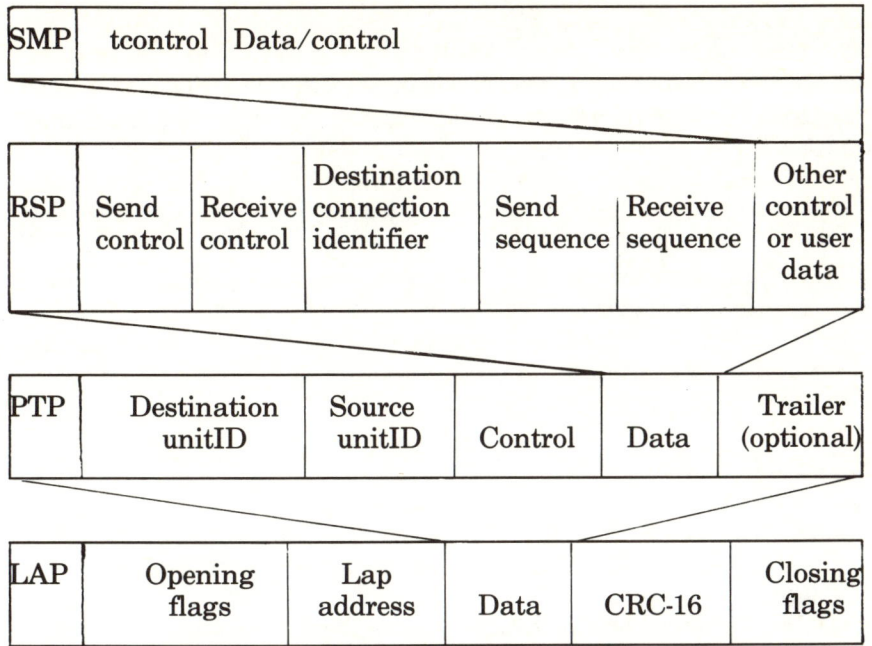

Fig. 15.11 LocalNet packet encapsulation.

CONCLUSIONS: IMPACT OF BROADBAND ON PROTOCOL DESIGN

As might be expected, the impact of broadband on the design of the protocols is more noticeable at the lower than at the higher levels.

In particular, the link level LAP protocol is very much tailored to broadband. The use of HDLC framing allows end-of-frame to be easily recognised by a receiving station even if the transmitting broadband modem has a shut-off time of several bits. Similarly, the technique used for collision detection is optimised for the broadband channels. Baseband oriented collision detection techniques which assume that a station can simultaneously listen to each bit as it is transmitted simply will not work on broadband, with its inherent signal delays through the headend. Consequently, the LocalNet link level protocol uses the 'byte comparison' approach to collision detection, which allows transmitted bits to be stored for later comparison purposes as they are received.

At the network level, the multi-channel broadband environment dictates a flat address space, since the presence of frequency-agile modems does not allow the channel number to become a permanent part of a station's address. Given this frequency-agile situation, the broadcast capability of the broadband channels is used to advantage in the network-level 'discovery' operation via which nodes may be easily located no matter what channel (or cable system) they happen currently to be on.

The transport level protocol RSP is tailored for efficient implementation on simple microprocessors; however, it provides full virtual connection services and is not explicitly oriented towards broadband. Similarly, the design of the LocalNet session level and presentation level services was driven more by the desire to support a wide variety of devices and applications than by the requirements of an underlying broadband environment.

REFERENCES

1. *Basic Reference Model for Open Systems Interconnection* (1981) ISO/TC97/SC16 DP7498, International Standards Organization, August.
2. *DCEC Protocol Standardization Program – Proposed Transmission Control Protocol Standard* (1982) SDC TM-7172/301/00, Systems Development Corporation, July.
3. *Internet Transport Protocols* (1981) Xerox Network Systems Integration Standard, XSIS 028112, December.
4. Tanenbaum, Andrew S. *Computer Networks* (1981) Prentice-Hall Inc., Englewood Cliffs, NJ, 148–164.
5. Metcalfe, R. M., and Boggs, D. R. 'Ethernet: Distributed Packet Switching for Local Computer Networks' (1976) *Comm. ACM*, **19**, 395–404, July.
6. *Standard ECMA-72 Transport Protocol* (1982) European Computer Manufacturers Association, September.
7. Day, John D. 'Terminal Protocols' (1980) *IEEE Trans.* **COM-28**, 4, April.

CHAPTER 16

LocalNet 20 Modelling Analysis and Performance

TRAFFIC MODELLING AND CHANNEL LOADING ANALYSIS

In the previous chapter we have discussed the protocol architecture that implements a particular broadband local area network product – Sytek's LocalNet.

Figure 16.1 defines the LocalNet packet encapsulation to implement the protocol architecture. The discussion that follows will use the above definition to estimate the allowable channel utilisation and delays within such a network service.

From Fig. 16.1 we see that each of the protocol layers contains the following header/trailer bytes:

Session (SMP):   1
Transport (RSP): 5
Network (PTP):   5 excluding optional trailer
Link (LAP):   3

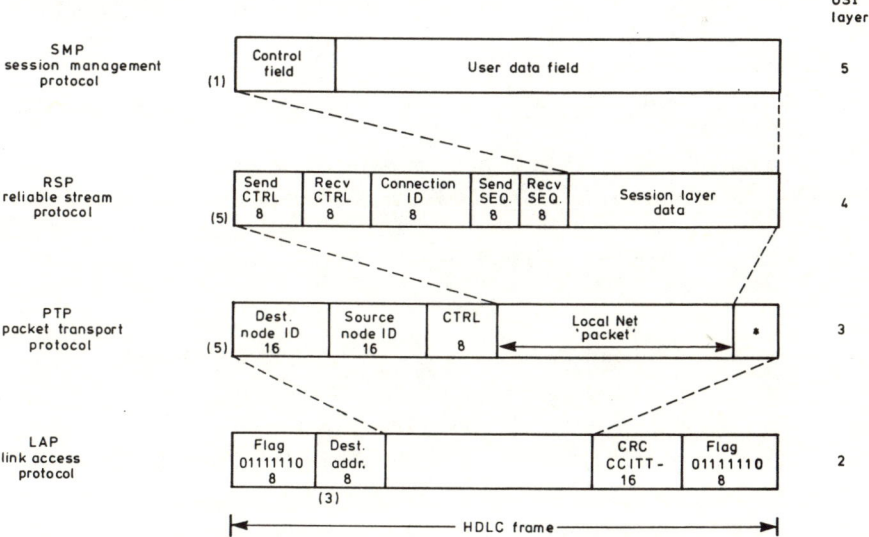

* - Optional trailer field (5 octects)

Fig. 16.1 LocalNet layered protocol formats.

Therefore a packet carrying user data across the network will consist of 14 bytes plus user data bytes. A packet used for control only would be generated at the transport level and therefore only consist of 13 bytes. The above ignores the flag characters that are used to delimit the start and end of packets.

A minimum of four addition flag characters are transmitted to assist the receiving modem to synchronise itself correctly to receive the following data and control characters. LocalNet also uses additional flag characters as part of the collision detection mechanism to ensure that the channel is free before data or control packets are transmitted. Additional flag characters are transmitted ahead of data and control packets, the number of characters transmitted beeing proportional to the distance the transmitting node is located from the headend (see Fig. 16.2).

The maximum radial distance travelled by 1 byte at 128 kbps with coaxial cable is given by

$$R = (1 \text{ byte}) \times (8 \text{ bits}) \times (\tfrac{1}{128} \text{ kbps}) \times (\text{velocity of propagation}) /_2$$

where $R \approx 5.8$ miles or 9.378 km.

To detect collisions the sending node, while transmitting flag characters, continues to listen to the receive frequency and checks that the received characters are also flags. If the transmitted and received characters match then no collision has occurred, the sending node stops listening to the channel, and the remainder of the frame is transmitted. If the characters do not match then it is assumed that a collision has taken place on the channel. The sending node stops transmitting and the frame is rescheduled for retransmission at a later time.

The number of flag characters transmitted ahead of the LocalNet packet determines the CSMA/CD network slot time or maximum

CRF	
miles	
5	Basic packet +1 flags
10	Basic packet +2 flags
15	Basic packet +3 flags
20	Basic packet +4 flags
25	Basic packet +5 flags
30	Basic packet +6 flags
35	Basic packet +7 flags

Fig. 16.2 Length of packets *vs*. cable distance.

operating radius. The minimum of four additional flag characters will give a network radius R of:

$$R \approx 5.8 \times 4 = 23.2 \text{ miles or } 37.5 \text{ km of cable}$$

Figure 16.3 illustrates how collisions occur between packets transmitted by a number of communicating nodes. If Node 1 sends a sequence of flag characters out on the channel, before it reaches Node 2 Node 2 begins transmitting its own flag characters, since it had not yet detected the flat characters of Node 1. The number of flag characters transmitted by Node 1 is such that a collision would have occurred and had time to travel back to Node 1 while Node 1 was still transmitting the initial flag character sequence.

Node 1 therefore associates the collision with the packet it is attempting to transmit. Collisions are detected before packets are transmitted. If flag characters are received correctly by the transmitting node, the transmitting node stops listening to the channel and

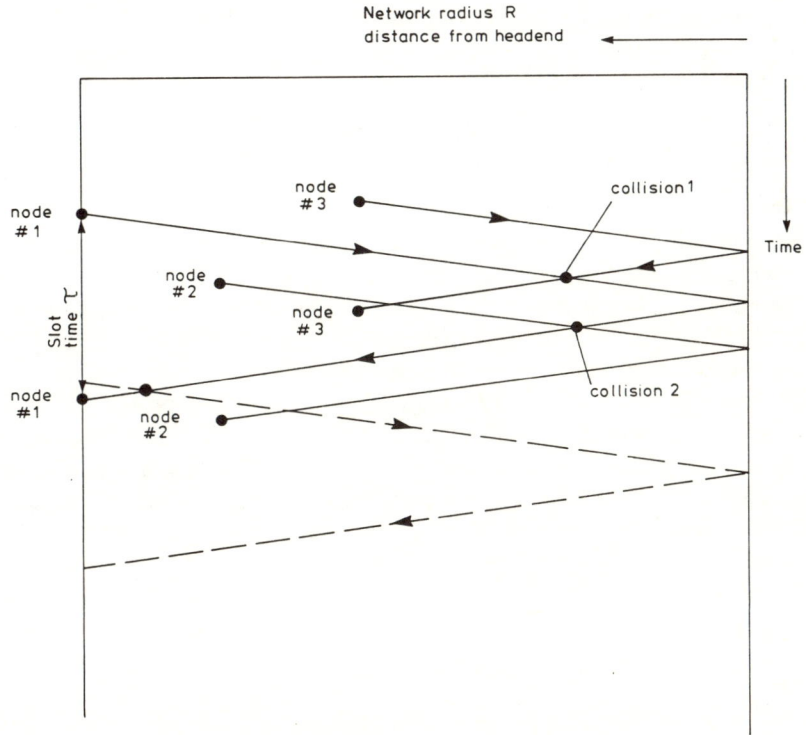

Fig. 16.3 Propagation delays and collisions.

transmits the remainder of the packet. CSMA/CD link level protocol is designed to ensure best effort delivery of packets.

In networks that use CSMA/CD as a media access protocol there is a trade-off between the maximum operating radius R (metres) and the transmission speed S kbps for a given minimum packet size, N (bits):

$$R \times S = N \times C \times \frac{1}{2}$$

where C = velocity of propagation (m/s).

Therefore for a given minimum packet size:

$$R \times S = \text{constant}$$

If we increase the transmission speed then for the collision to be detected within the transmission time of the minimum packets the maximum radius of the network must be reduced. If $N = 64 \times 8$ bits, $S = 10 \times 10^6$ bits/s, and $C = 0.7 \times 3 \times 10^8$ m/s, then:

$$R = \frac{64 \times 8 \times 0.7 \times 3 \times 10^8}{10 \times 10^6 \times 2}$$
$$R = 5376 \text{ metres}$$

In practice, additional delays are introduced by the head and translator equipment; $1\,\mu$ is equivalent to a reduction in the radius of the network by:

$$0.7 \times 3 \times 10^8 \times 1 \times 10^{-6} \text{ m}$$
$$= 2.1 \times 10^2 \text{ m}$$
$$= 210 \text{ metres}$$

The use of CSMA/CD as a media access protocol is an ideal engineering compromise to support slow speed terminal traffic over a large geographic area. A data channel running at low data rates (e.g. 128 kbps) and hence requiring low bandwidth would support many low speed terminals over large lengths (20 miles) of cable.

The use of CSMA/CD over a high speed (10 Mbps) channel would require a proportional reduction in cable length of 3 miles. In applications when communication is required at high speed over large distances of cable, an alternative *token bus* media access protocol should be considered.

ALLOWABLE CHANNEL UTILISATION AND ACCESS DELAY

Since the broadband channel is shared by many nodes by means of the CSMA/CD link level protocol we need guidelines for system designers for the following network parameters:

(1) *Average channel loading.* How heavily can a channel be loaded?
(2) *Average access delay.* What is the average delay at recommended channel loading before a transmitting node can access a channel?

Assuming the RF distribution cable is operating to specification we may apply the results of analysis carried out by Tobagi and Hunt in their paper, 'Performance Analysis of Carrier Sense Multiple Access with Collision Detection' (Fouad A. Tobagi and V. Bruce Hunt, Computer Systems Laboratory, Standford University, California, Proceedings of the LACN Symposium, May, 1979).

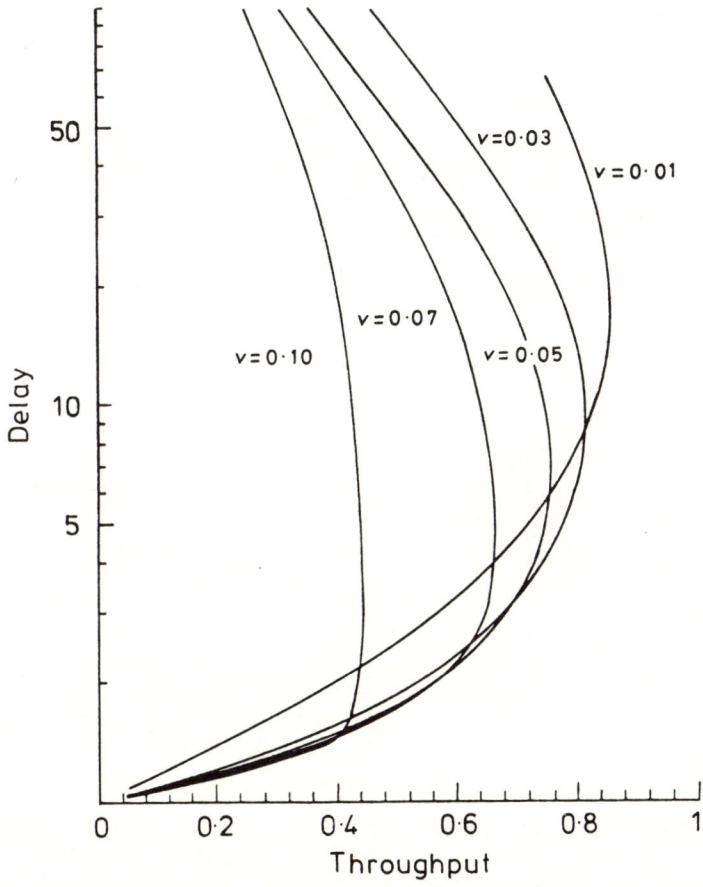

Fig. 16.4 The throughput-delay trade-off in CSMA at fixed v.

The analysis carried out by Tobagi and Hunt is summarised in Figs. 16.4 and 16.5. These graphs illustrate the throughput – delay tradeoff in CSMA/CD at fixed v. Where γ is defined as $1/_\gamma$ = Mean rescheduling delay of backlogged packet (slots).

The graphs illustrate the fact that as the throughput approaches 90%, delay increases dramatically. It is therefore prudent when designing systems to keep the traffic on a channel to less than 70%. Below 70% channel loading the average access delay is approximately proportional to the channel loading or throughput.

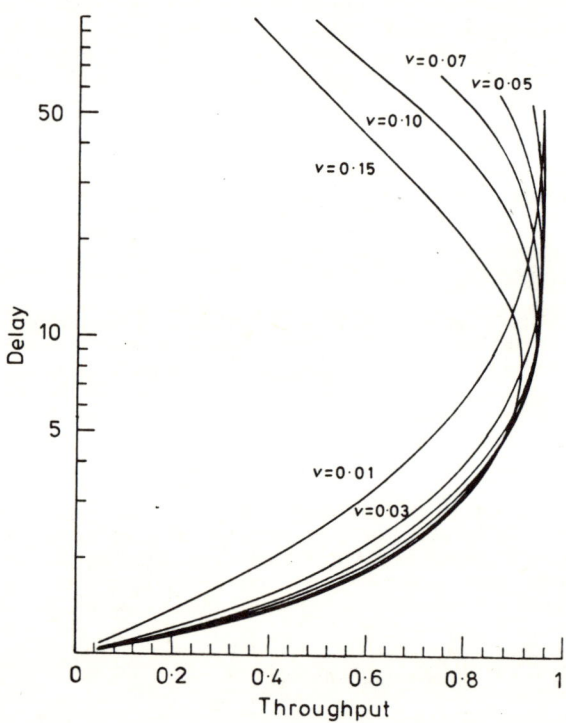

v = 2
τ = unit time
τ = slot time
Rescheduling delay mean = 1/v slots
M = 50 devices
T = 100 (normalised transmission time of a packet)
γ = 2

Fig. 16.5 The throughput-delay trade-off in CSMA/CD at fixed v.

Using the graphs published by Tobagi and Hunt we can estimate the average access delay introduced by CSMA/CD for a Sytek LocalNet channel:

CSMA/CD delay td = (from graph at 0.7 throughout) $\times$ tnet

where tnet = time for packet to traverse network
$$\text{tnet} = \frac{\text{bits/packet}}{128\ 000\ \text{bits/s}}$$
For 70% loading:

$$td = 5 \times \text{tnet}$$
$$td = 5 \times \frac{\text{bits/packet}}{128\ 000}$$

Therefore the additional access delay for a 70% loaded LocalNet channel with the following packet characteristics are:

No. of data bytes in packet = 30
Cable radius   = 5 miles
$$td = (5)\left(\frac{30(8) + 14\ (8) + 8}{128\ 000}\right)$$
$$td = 15\ \text{ms @ 70\% channel loading}$$

Many existing applications for packet switching consist of connecting asynchronous terminals to host minicomputers. These simple asynchronous terminals operate so that for every key depression a character is transmitted to the minicomputer, and the minicomputer accepts this character and then echoes it back to the terminal for display to the user. This mode of operation is known as *echoplexing*. When operating in this mode each packet will consist of 1 data byte plus the normal header/trailer characters associated with the packet protocols. The bits per packet for this application can therefore be computed and the total access delay for a 70% loaded LocalNet channel estimated:

$$td = (5)\left(\frac{1(8) + 14(8) + 4\ (8) + 8}{128\,000}\right)2$$
$$td = 12.5\ \text{ms @ 70\% channel loading}$$

A user at an asynchronous terminal will not detect this short additional delay – 150 ms is generally considered the threshold of human visual perception.

Note that in addition to the above access delay are the normal computing delays within the packet communication units executing the higher level protocols.

MODELLING EXAMPLE

Once the elements of the data communication protocol have been defined together with the data packet encoding it, it is then possible to model a particular transaction sequence. Illustrated below is a transaction sequence between an asynchronous terminal and a minicomputer, and the modelling example is used to estimate the number of terminals that could be supported by a single data channel.

Assumptions: 5 cable radius
       Transaction network:
        6 characters echoplexed
        2000 characters back
       2 transactions per minute per user
       70% channel utilisation
       2:1 port contention ratio
       Essentially half-duplex
       Non-bridged session

Transaction model:

 Echoplex: CHAR
         LN ACK
         CHAR
         LN ACK

Character packets (2)
bits $= 1 (8) + 14 (8) + 4 (8) + 8$
 $= 160$ bits

ACK packets (2)
bits $= 13 (8) + 4 (8) + 8$
 $= 144$ bits

Single echoplexed character
bits $= (160) 2 + (144) 2$
 $= 608$ bits

2000 character screen return
Number of packets:

$$\frac{2000}{64} = 31 +$$

$\therefore$ 31 – 64 data
 1 – 16 data

64 data packets (31)
bits $= 64 (8) + 14 (8) + 4 (8) + 8$
  $= 664$ bits

16 data packet (1)
bits $= 16 (8) + 14 (8) + 4 (8) + 8$
  $= 280$ bits

ACK = 144 bits
Screen
bits = 664 (31) + 280 (1) + 144 (32)
       = 20 584 + 280 = 4608
       = 25 472
Transaction total
  bits = 6 (EP) + 1 (screen)
       = 6 (608) + 1 (25 472)
       = 3648 + 25 472
       = 29 120 bits
Calculate transactions per minute.
Transactions/channel/min

$$= \frac{\text{(Channel speed) (Available percentage)}}{\text{Transaction data quantity (bits)/m/s minutes--seconds conversion}}$$

$$= \frac{(128\,000)\,(0.70)}{(29\,120)/(1/60 \text{ m/s})}$$

= 184 transaction/min
Number of active devices/channel

$$= \frac{\text{Number transactions/min}}{\text{Transactions/device/min}}$$

$$= \frac{184}{2}$$

= 91 active sessions
With 2:1 port contention
= 184 user devices/channel
Comments: data efficiency low due to high number of packets for
      echoplexing
Efficiency: $\dfrac{16\,096 \text{ (data bits)}}{29\,120 \text{ (total bits)}}$
              = 55% efficiency
Bridges and links:
   Bridges between channels naturally slow down windowing open-
   ing by having outstanding packets in store and forward memory
   Echoplex delay for CSMA/CD
      CSMA/CD channel access time contribution to packet delay will
      include both cable access times
   Any packet which is bridged will be existent on both channels
Methods of improving performance:
   Increase number of channels, lower utilisation
   Group users and resources
   Allow frequency (channel) agility

PERFORMANCE MEASUREMENTS

The measurement of network performance, especially when all the processing elements are totally distributed, is difficult to set up and control if it also includes a large number of transactions across a number of nodes.

Simple experiments are, however, easy to set up in the laboratory and do give an indication of the processing delays that result with high level packet protocols.

Two types of measurements can be undertaken:

(1) Added time delay due to protocol processing.
(2) The steady state throughput.

Knowledge of the additional time delay is useful when the network forms part of a transaction processing system. The steady state throughput will affect the speed of file transfers across the network.

Experiments have been undertaken to measure the added delay due to protocol processing in LocalNet 20 packet communication units.

Added delay - single character echoplex

Using a data communication protocol tester (e.g. Atlantic Research 4500) one can determine a baseline time for transmitting a character and then echoing it between two Atlantic Research 4500 protocol testers that are directly connected to each other with a piece of wire.

The above experiment can now be repeated with the data being transmitted between two packet communications units across an unloaded broadband data channel. The difference in time between the two experiments is the added delay.

From the result shown in Fig. 16.6 we may draw the following conclusions:

The protocol processing time per character per node is 15 ms.

The additional delay when transporting a character via a bridge is 20 ms.

ADDED DELAY - SCREEN TRANSFER

The next experiment measures the added delay when relatively large blocks (VDU screens) of data are transmitted across the network.

This time the experiment is repeated between two protocol testers such that a single character transmitted by one of the protocol testers

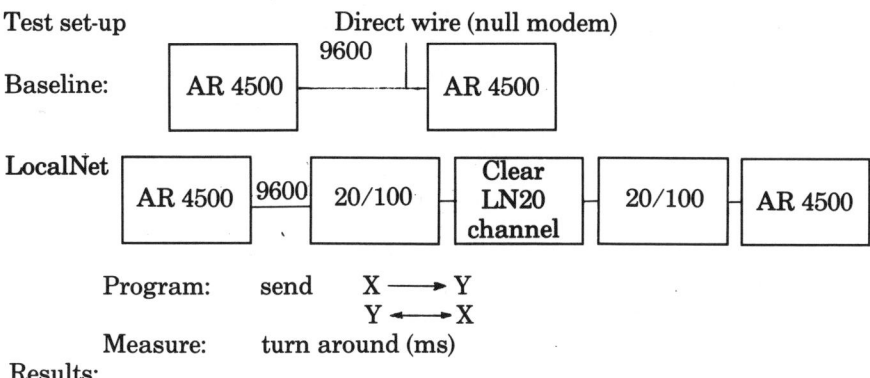

Test set-up   Direct wire (null modem)

Program: send   X ⟶ Y
                              Y ⟵ X
Measure: turn around (ms)

Results:

Baseline: 4.6 ms
LocalNet 20/100.

Note: baseline time has been subtracted from 20/100 to show added delay

PCU status		V2 software		V2.2 software	
EOM	Idle	No bridge	Bridge 2.0	No bridge	Bridge
0	50 ms	127 ms	140 ms	112 ms	129 ms
0	10 ms	37 ms	63 ms	29 ms	55 ms
1 char.	0	30 ms	46 ms	31 ms	41 ms

Idle = idle timer.
EOM = end of message.

Fig. 16.6 20/100 delay measurements, single character echoplex.

will cause the other to transmit a simulated VDU screen of data in response. The experiment is carried out when the protocol testers are directly wired to each other and the turn-around time noted.

The experiment is then repeated between the two packet communications units across an unloaded broadband channel. The difference between the two experiments once again is the added delay due to the packet protocols.

Figure 16.7 summarises the result from such an experiment. The additional protocol processing time is approximated to 150 ms. A screen full of data would usually be segmented in 64 character buffers before being transmitted. This would result in at least 31 buffers being transmitted. If the delays were attributed evenly to the

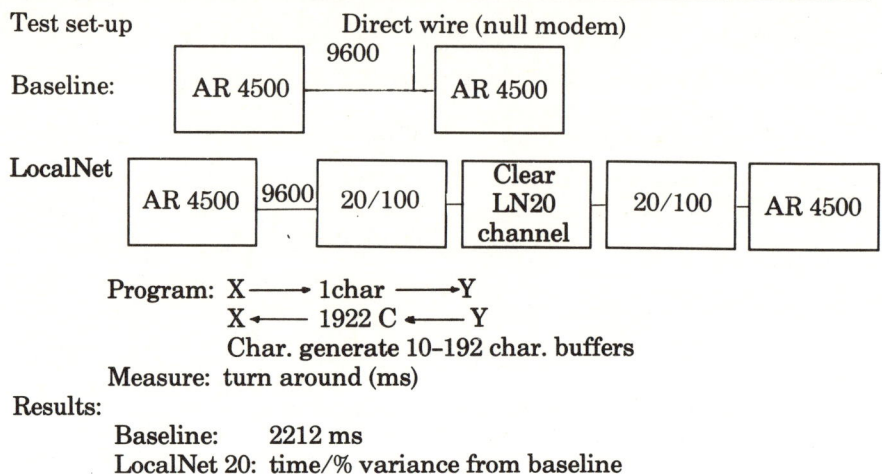

Program: X ———→ 1char ———→Y
         X ←——— 1922 C ←——— Y
         Char. generate 10–192 char. buffers
Measure: turn around (ms)

Results:
    Baseline: 2212 ms
    LocalNet 20: time/% variance from baseline

Test system	PCU software			
	V2 → V2	V2 → V2.2	V2.2 → V2	V2.2 →V2.2
No bridge	2360/6.7%	2355/6.4%	2354/6.4%	2342/5.9%
Bridge W/2.1	2406/8.8%	2368/7.1%	2353/6.4%	2362/6.8%
Bridge W/2.0	2822/27.6%	2371/7.2%	2377/7.5%	2370/7.1%

Fig. 16.7 20/100 added delay measurements, screen return.

computation required to segment and transmit each buffer then the additional time per buffer would be approximately 5 ms.

The results obtained from the two experiments confirm the fact that packet protocols are more efficient when transmitting relatively large buffers of data and are inefficient at transmitting single characters across a network.

STEADY STATE DATA TRANSFER

Finally, Fig. 16.8 illustrates an experiment to determine the effect of packet protocol on the steady state of data transfer. Packet protocols must be implemented so that there is adequate computing resource to achieve steady state data transfer at a particular speed.

Test set-up

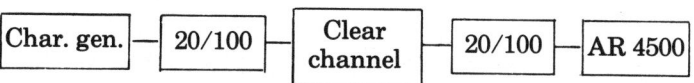

10 bit data at 9600 baud

Results: bits per second/% of 9600 bps

Test system	PCU software			
	V2 → V2	V2 → V2.2	V2.2 → V2	V2.2 → V2.2
No bridge	8616/89.8%	9576/99.8%	9464/98.6%	9576/99.8%
Bridge W/2.1	8368/87.2%	9224/96.1%	9160/95.4%	9576/99.8%
Bridge W/2.0	7464/77.8%	8984/93.6%	9160/95.4%	9576/99.8%

Fig. 16.8 20/100 steady state data transfer.

This experiment measures the effect of steady state data transfer at 9600 bps using various versions of software. It can readily be seen that the later versions of software are more efficient.

CHAPTER 17

The Use of Broadband for the Distribution of and Access to Financial Information Services

Banking and finance have become information based industries – 'information means power and prior information means wealth'. Today, when smaller banks may trade sums equivalent to their whole balance sheets several times a day in the foreign exchange markets alone, speed and accuracy of information are even more crucial. Markets are more volatile: currencies and interest rates fluctuate tremendously. Financial instruments – bonds, shares, options, futures, swaps, etc. – are more numerous, varied and complex. One way or another, information is not only essential to the workings of a financial firm, but has become a financial product every bit as important as money. So systems for handling information are vital.

Successful trading in today's environment requires immediate access to a wide variety of information from:

Broker service displays.
Quotation systems.
News services.
Electronic trading systems.
Interactive systems.
In-house systems.
External time sharing systems.
Analytical systems.

Currently many of the services require their own visual display units (VDUs), keyboards and controllers. This results in a crowded maze of equipment on the dealer's desk. Operating such equipment can be complex, thus reducing the efficiency of the dealer. Efficiency is also reduced because people cannot easily or economically be relocated to other dealer positions because the necessary information is not available at the alternative location.

The requirement, therefore, is for an information distribution network in which the dealer or analyst has access to financial information and computing services using one streamlined device.

FINANCIAL INFORMATION SERVICES

There are a number of organisations that provide databases that are continuously updated with information from equity, gilts, commodity and foreign exchange markets. The list below are some of the leading financial information services:

Reuters Monitor.
Reuters Quotation.
Reuters Dealing.
Quotron.
Telerate.
Topic.
Datastream.
FBI.
GARBAN.
RMJ.
Bondspec.
Eurodata.
LIFFE.
Nasdag.
Seaq (Stock Exchange Automated Quote System).

The information is available to the user is in two formats:

Video.
Digital data.

VIDEO

A number of information services only present their information service via a terminal provided by the information service itself. The terminal provided by the information service consists of a controller plus a number of keyboards and VDUs. The controller is connected to the remote database computer system via leased telecommunication lines. A controller executes the communication protocol necessary to transfer data to and from the remote database computer system (Fig. 17.1). The keyboard is used to select the appropriate pages of information from the database or interact with the information service where appropriate. The controller outputs the information received from the remote database and displays the selected pages of information on a VDU. The electrical format of the signals used to drive the VDU is known as *composite video*. Many information services currently make available the service to the end user via a controller provided by the service, and the received information is

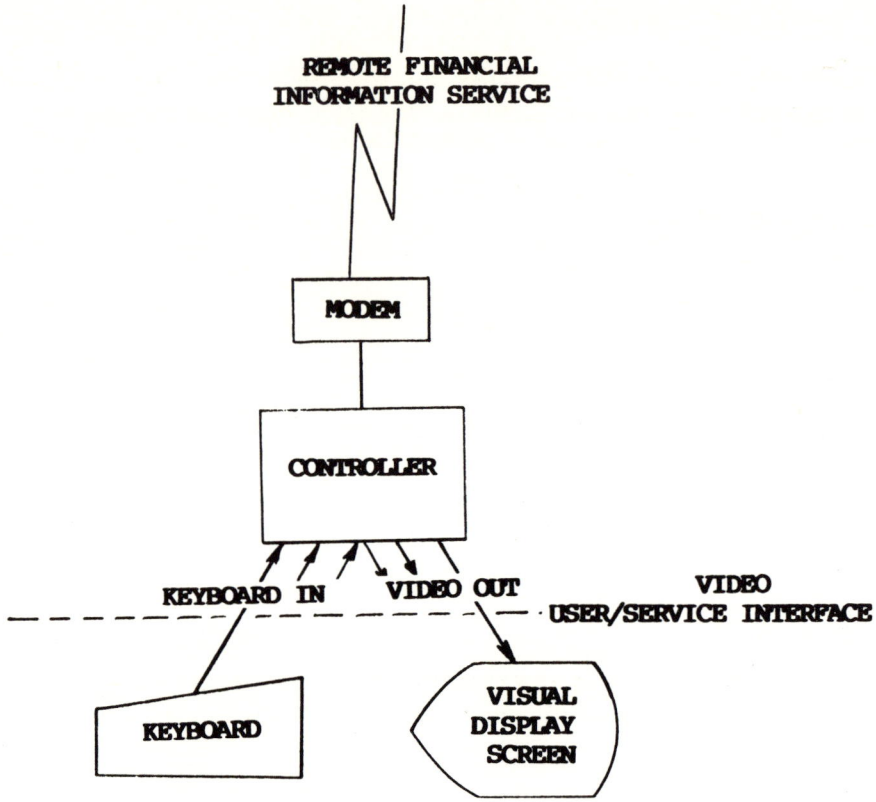

Fig. 17.1 Information service – video interface.

available only in video format. Information available in video format cannot be economically processed by a digital computing system. The advantages to the information provider in presenting a service with a video format are:

• The end user does not need to know any details of the communication protocol necessary to transfer data to and from the database.
• The copyright of the data is secure.

DIGITAL

Alternatively, the information service may provide access to the information in digital form, allowing the information to be accessed by a digital computer. The information received from the service can be input into a user's computer, stored, manipulated and redistributed to terminals attached to the digital computer (see Fig. 17.2).

The information service will provide pages from the database by allowing the end user to access the database computer directly. To make this method of access possible, the information service must define to the end user the details of the communication protocols used.

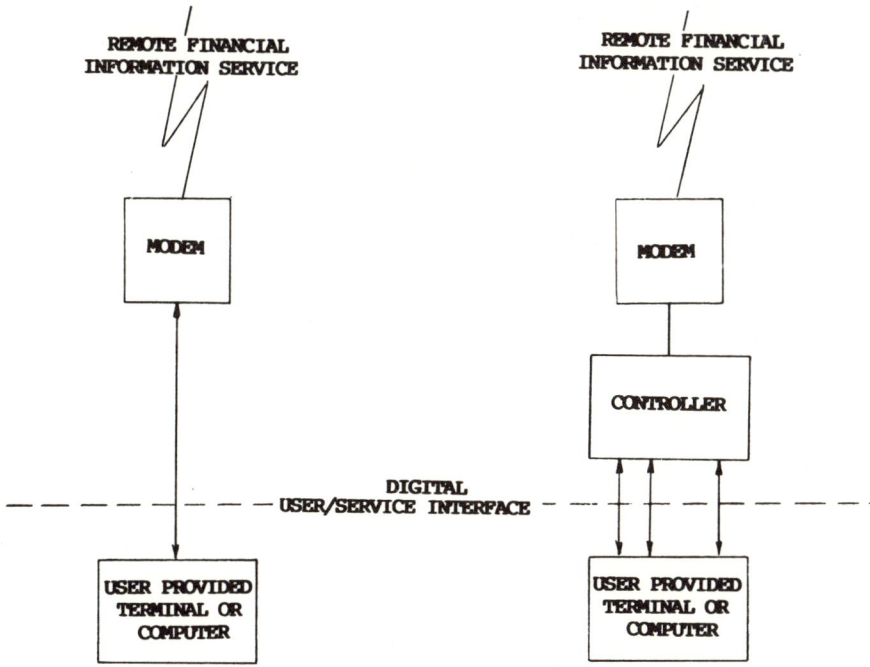

Fig. 17.2 Information service – digital interface.

Other information services do not allow user direct access to the database computer directly, but only via a controller provided by the service located at the customer's premises. Here the controller executes the more complex communication's protocol, and the user's terminal or computer equipment need only implement simpler access protocols to communicate with the local controller.

USER REQUIREMENTS

The user requires a practical solution to the proliferation of VDUs and keyboards. An information distribution network is required where via a single keyboard and one or more video screens, a dealer or analyst can call up and display any information on any screen. The system must be modular and not be restricted to a fixed number of input information sources or output displays. When a new service is added, it must become immediately available to all users.

A schematic showing the overall basic user requirement is depicted in Fig. 17.3. The user workstation consists of one or more video screens and a keyboard connected to a controller. The information distribution network must be capable of distributing both video and digital information. In addition, it must be capable of switching, under user command, both video and digital information from a variety of information sources to selected video screens.

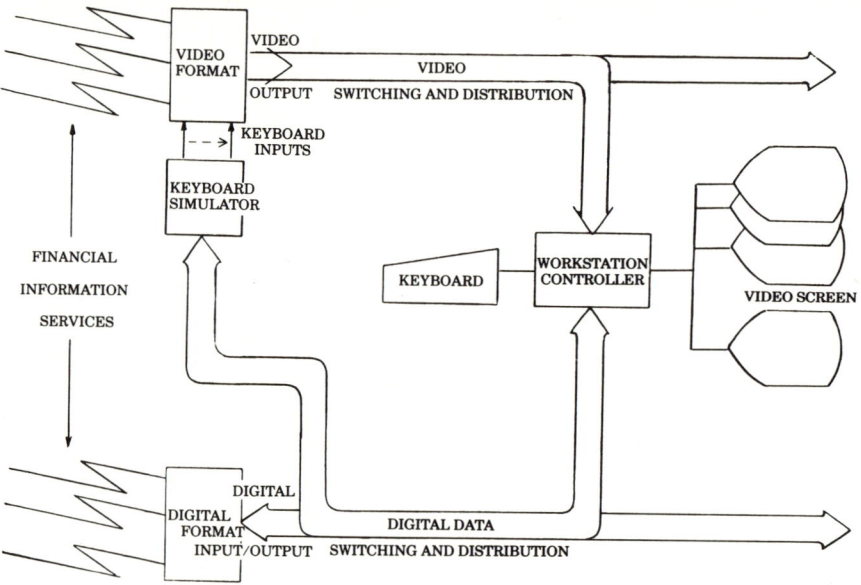

Fig. 17.3 Information distribution network, video plus digital data.

BROADBAND MEDIA

Broadband coaxial cable is an excellent medium for the distribution of both video and digital information. Broadband coaxial cable forms the basis of cable TV networks. Programme material consisting of many channels of television entertainment is broadcast over many kilometres of coaxial cable and accessed by thousands of subscribers.

Cable TV broadcast technology is well developed and equipment is readily available. Recent UK government legislation requires cable TV operators to implement *switched star* networks, which enable a wide range of interactive services to be provided via the cable TV network. As a result of these requirements, cable TV suppliers have developed subscriber controlled RF switches that will enable the subscriber to select a particular TV programme from about 30 programmes being broadcast on a single coaxial cable. The same cable technology may be used to broadcast financial information services that are currently only available in video format. The video ouput from the service is modulated onto a radio frequency (VHF) carrier in exactly the same way as an entertainment TV programme is modulated to occupy a channel on the cable. Up to 30 channels of financial information may be simultaneously broadcast over a single coaxial cable system. Many thousands of users may access any channel providing an information service via an RF switch and view the pages of information currently being broadcast.

On some coaxial cable transmission system, in another frequency channel, a digital data network for interactive data communications may be implemented. This network will allow data communications between terminals, workstations and host computers. This data communications network is generally known as a *broadband local area network*.

Broadband coaxial cable as a transmission medium has the capability of distributing both video and digital information via the same cable.

RF SWITCH

Information services that are currently only available in video format require to be distributed and accessed from a variety of locations within a building, across a campus site or from individual offices across a city.

In order to distribute the video outputs available from the information services, they are first modulated and then broadcast via the coaxial cable TV network. This is shown schematically in Fig. 17.4. User's video screens (TV sets) are connected to the services via what is commonly known in the cable TV industry as a *set top convertor*. The set top convertor is located at the user's desk, connected to the video screen (TV set), and interfaced to the RF switch which is located elsewhere, remote from the user. Information services are accessed by keying in the correct channel number to the set top either manually, via a workstation controller, or initiated remotely from a computer system. The channel number input will, if allowed by the network management system, cause the RF switch to select the appropriate channel and thereby deliver the selected information service to the user's video screen. This mechanism allows many thousands of authorised users to view information services currently being broadcast.

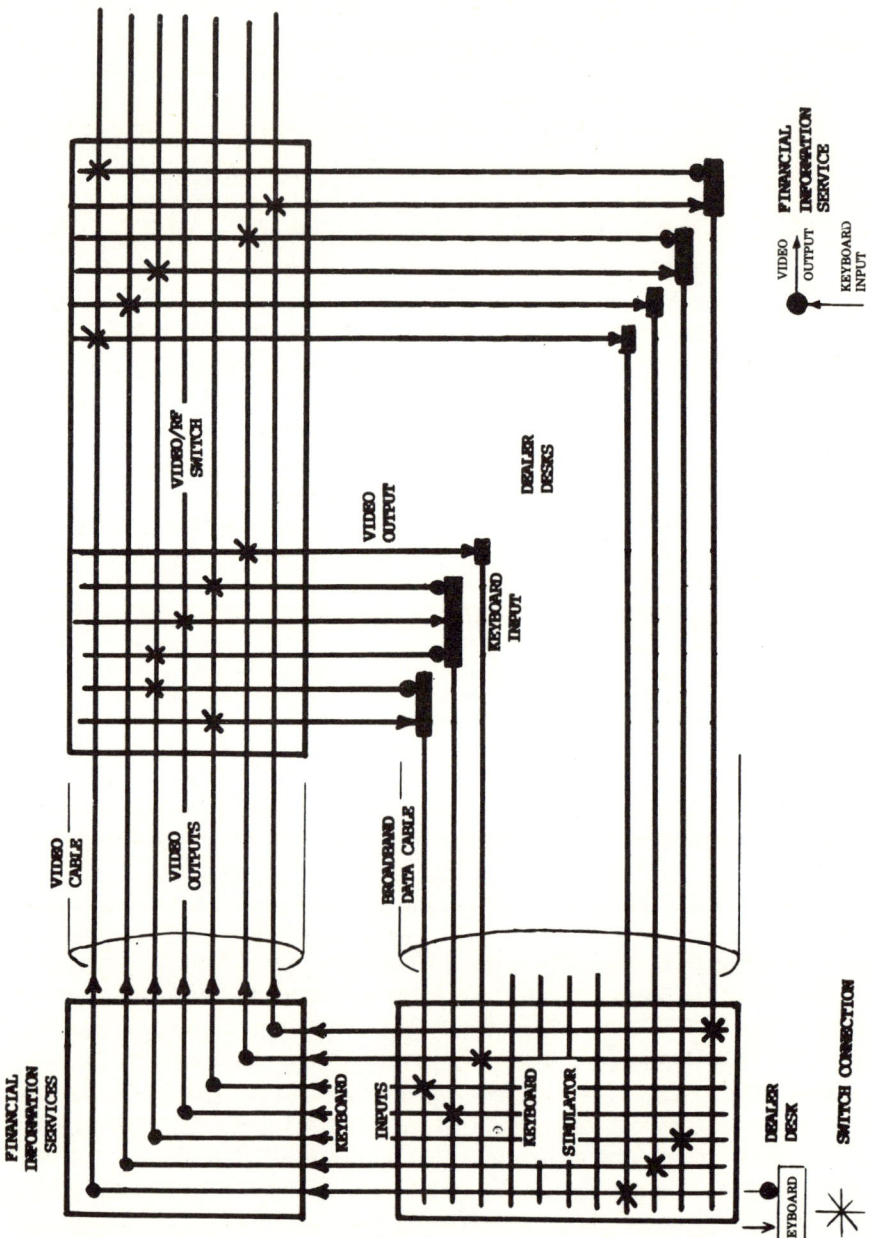

Fig. 17.4 Distribution of and access to financial information that is

DIGITAL INFORMATION SOURCES

Information services that are supplied in digital form may be presented to the user in the following ways:

- A terminal supplied by the information service.
- A communications controller supplied by the information service to which the user attaches a terminal or computer system.
- A leased or dial-up telecommunication line to the information service, plus a definition of the communications protocol used by the service.

Where an organisation subscribes to many information services, large numbers of users need to view many pages from information sources simultaneously. Users need these multiple information sources but cannot afford the space or expense of installing a keyboard and display per information source, per user.

Digital data feeds, now increasingly available from the major information vendors, will allow these sources of information to be taken into a central in-house digital computer for reformatting and re-distribution. Users now will only require a single keyboard and one or more video screens. The user can then interact with the in-house computer to select and view information received from several information sources.

KEYBOARD INPUTS

In addition to viewing pages of information, users need to interact with the information service to select other pages of information from the same source or to input data to the service. Associated with each information source video ouput is a keyboard input; data and commands input will be accepted and processed by the remotely located computer system at the information provider.

From Fig. 17.4 we see that there may be thousands of users viewing pages broadcast from a limited number of information sources. Only user keyboards that are currently connected to the keyboard inputs via the keyboard simulator may interact directly with the information sources. The maximum number of users that may simultaneously interact with the information sources equals the number of keyboard inputs available. User keyboards, therefore, contend for the limited number of keyboard inputs available. Contention for keyboard inputs will be provided by the broadband local area network.

MULTIPLE VIDEO SCREENS

A user can have more than one video screen. Any page available in video format from the information service may be displayed on any video screen selected by the user. Only the users whose keyboards are connected to the available keyboard inputs may interact directly with the information source. All other users may only view the pages selected by the users that are on line. If no data is passed from the user's keyboard that is on line for a period of time, the broadband local area network can disconnect the user's keyboard from the keyboard input. This keyboard input is now available for use by other users.

Pages received from information sources can be stored by the digital computer and then re-distributed to users on demand. Each user screen can display data from one source at a time and the data will be in a format determined by the information vendor or contributor, not by the recipient. Thus there is still a problem for the user in the assimilation and comprehension of the information.

In order to aid in the assimilation and comprehension of information, users need to be able to define personal screen formats which contain data from a variety of sources, presented according to their own needs.

A foreign exchange dealer may require on a single screen:

- Currency position.
- Customer's current limit and exposure.
- Currency rates from Reuters and Telerate.

In the equities market, the research department's data and the financial analyst's opinion could be displayed simultaneously, with historical information from Datastream and current price data from the Stock Exchange Hotline feed.

By constructing personalised screen formats from multiple sources of information, the need to display simultaneously raw information from several sources on multiple screens is reduced.

Composite pages of information can be formatted at the central in-house computer that is handling the communications with the information sources, or the formatting can be done on a distributed basis at the user's intelligent workstation.

Users of information services frequently require to access and view pages of information from a number of souces. A centralised in-house computer that supports a large number of terminals and/or data feeds would be responsible for all processing for communications, formatting and display. The larger the number of terminals and higher the rate of request to display pages, the larger is the processing and input/output load on the computer. Therefore the response time will lengthen when the number of terminals attached to the computer or the number of page requests increases.

In order to reduce the processing load, formatting and displaying pages can be carried out on a distributed basis at the user's workstation, provided the information sources received by the central computer can be distributed to a large workstation population without significantly increasing the processing load.

Information can either be provided on request to the terminal or workstation, broadcast to all workstations only when a change occurs, or by broadcast of all pages at frequent intervals. The first of these methods gives acceptable response time for a small number of terminals with low rates of request for the display of pages. For large numbers of terminals with high rates of display page request, the response times become unacceptably long. The second method requires that information is stored at the workstations and makes unsynchronised workstation start-up from power up or failure complex and slow.

Regular broadcasting of all information is the best approach, provided the broadcasting method does not increase the processing load significantly as the terminal population or the rate of request for pages increases.

BROADBAND INFORMATION DISTRIBUTION NETWORK

Broadband techniques enable us to combine into a single, coherent information distribution network the capability to:

- Distribute information presented in video format.
- Switch and select a particular source of information from several available in video format.
- Interact with video information source via a keyboard.
- Support several video screens at a user location for the simultaneous display of several information sources.
- Provide simultaneously on the same coaxial cable a packet switched data network service (local area network, LAN).
- Use the packet switched service to enable users via keyboards to access and interact with information services that are available in either video or digital format.
- Support data communications between user terminals/workstations and information sources available in digital format.
- Support the broadcast of large volumes of data on several channels via full field teletext.

TELETEXT BROADCASTING SYSTEM

Teletext is digital text and graphic information pages that are continuously broadcast by an information source (e.g. BBC Ceefax,

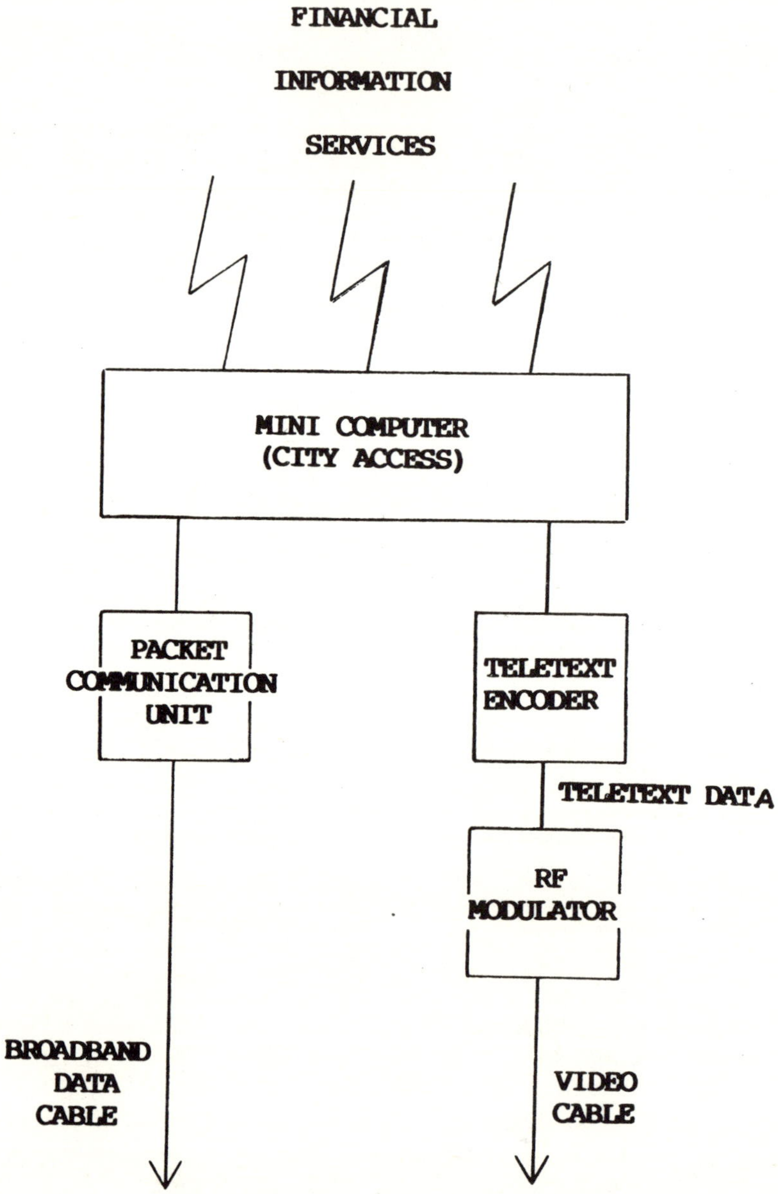

Fig. 17.5 Teletext broadcasting system.

ITV Oracle or Cable TV Operator). A television subscriber can select a particular information page from the pages being continuously broadcast and display this information on the television screen. Many thousands of subscribers may access the information being broadcast without affectinig the response times. (See Fig. 17.5.)

Conventional (VBI) Teletext

VBI or *vertical blanking interval* teletext encodes digital data onto portions of the television signal that are normally not viewed as part of the vision transmission. The information thus encoded may be inserted into an existing vision programme channel and broadcast. Since the portion of the television signal available for data is limited, VBI Teletext is used where the total number of pages to be broadcast is typically 100–200 pages and the access time is not critical.

Full Field Teletext

Full Field Teletext uses a whole TV channel for digital data transmission, no visual information being broadcast. A data rate of approximately 7 Mbps can be achieved. A page of information consisting of $24 \times 40 = 960$ characters may be transmitted at a rate of approximately 900 pages per second.

Figure 17.6 illustrates the elements of a broadband information distribution network. The system accepts information sources from external services that are presented in video and/or digital format. The video output available from communication controllers provided by the information services is modulated to occupy a bandwidth equivalent to a television channel and then broadcast via the coaxial cable distribution network. A single coaxial cable has the capacity to distribute approximately 30 different information sources to many thousands of video screens. Users may select and view any particular service on a video screen. By installing more than one screen at a user's location the user can then view a number of information sources simultaneously. Interaction between a user and the information service is via a keyboard. Where the information service is only available in video format only one user at a time can interact with the service. Users, therefore, will contend for the keyboard inputs available, the contention mechanism being provided by the packet switched network.

Information sources available in digital format are input into a digital computer. Users wishing to access a service now make an enquiry for the information at the digital computer. If the informa-

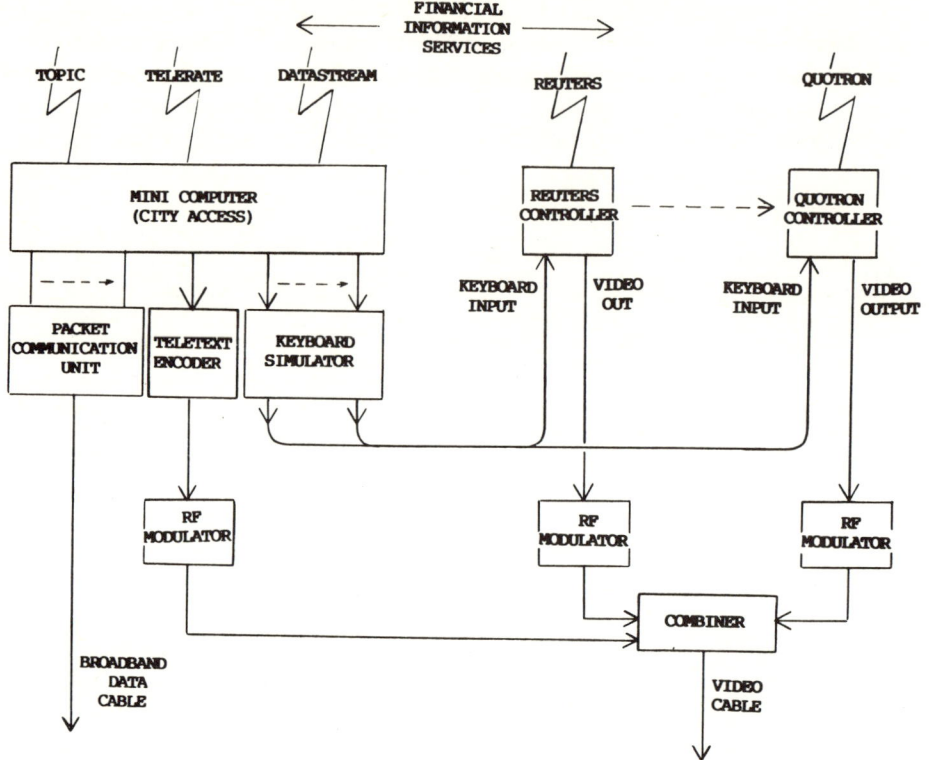

Fig. 17.6 Broadband information distribution network, video plus teletext
broadcast and digital data.

tion already exists, because another user had previously requested it,
no access is required on the information source; the digital computer
simply distributes information already held in its file store. If the
information does not exist in the file store, then the computer will
access the information service and obtain the information from the
remote database, distribute it to the user and store a copy in its file
store.

A Full Field Teletext broadcasting system enables many hundreds
of pages of information to be broadcast continuously at high speed.
Selected pages may be accessed using a Full Field Teletext decoder,
the information being displayed on a television set or a video
monitor. The average access time to a page of information depends
upon the number of pages being broadcast and not upon the number
of terminals or the rate of request for a new page to be displayed.

Access to the digital computer from user terminals/workstations is
via a packet switched service. This provides the user with switched
circuits for data communications, normally between the user termi-

nals/workstations and digital computing resources. One of the digital computing resources would be a communications processor to which are connected the communications lines from the remote database. Other digital computing resources required to be accessed by the user from his terminal/workstation may be used for specific transaction processing applications, which may include using data retrieved from the information sources.

When large numbers of users require frequently to view information derived from remote databases it is more efficient to broadcast a set of information pages at high speed over a network and allow users to select particular pages from the set of pages being continuously broadcast.

The technique used to broadcast pages of information – teletext – is well known. Users accessing the digital computer will cause a set of 'most frequently accessed pages' to be continuously broadcast. A user will, therefore, initially access the teletext broadcast pages for the necessary information. If the information is currently available the user will view the page without requiring to access the digital computer and therefore impose no transaction processing load on the computer. If the information is not available via the teletext broadcast pages then the user will need to access the digital computer and request that the appropriate page be added to the broadcast, replacing a page that is infrequently viewed.

Figure 17.7 illustrates a broadband information distribution network implemented on two coaxial cables. One cable distributes all the broadcast based information services, including all services that are available in video format plus pages distributed as Full Field Teletext. The second cable provides the packet switching network. In the example, two networks are shown – LocalNet 20 and IBM PC-NET. LocalNet 20 will be used when the user has asynchronous terminals, and IBM PC-NET where the user is using an IBM PC workstation.

USER WORKSTATION

In order to improve the user's ability to assimilate and comprehend information embedded in vast quantities of data, it becomes necessary for the user to be equipped with distributed intelligent workstations. The workstations will have the ability to select and accept data from information sources currently only available in video format and display the selected pages on a video screen. Because this data is in video format it cannot economically be processed digitally. Raw information available in digital form can be accepted by the user

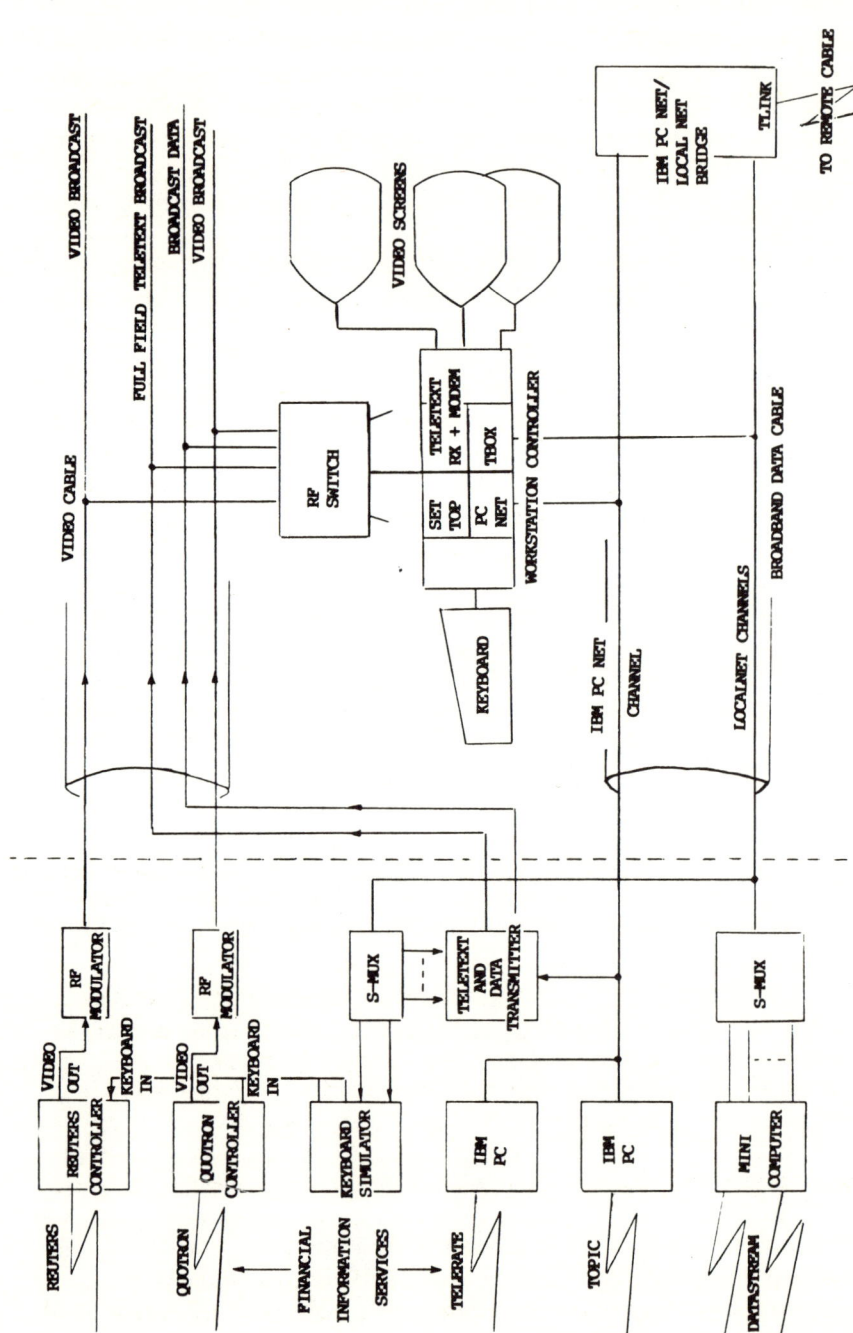

Fig. 17.7 Distribution of and access to financial information services.

workstation, re-formatted and displayed in a form that is determined by the user's own needs. Frequently used information pages will be available via the Full Field Teletext broadcast system and where special interaction is necessary with the communications processor this will take place via the packet switched channel. When appropriate, any pages built specifically for an individual user can, if desired, be transmitted from the user workstatioin to the teletext broadcast system and thereby made available to all other users of the information distribution network.

Thus, with datafeeds in digital form, personal computing and broadband communications technology, the means exist to provide users with information tailored to their specific requirements. To complete the picture the user must be provided with methods to define the content and layout of pages which he wishes to view. The requirements of various markets will tend to suggest different approaches.

In the foreign exchange and money market areas, the set of information and layout of pages is likely to remain relatively static, but the actual data (i.e. exchange rates) will change rapidly. In such circumstances, the user will only rarely wish to change the contents or layout and an off-line update of the page format would be appropriate.

If particular pages are likely to be used by a number of users and therefore central control of the formats is desirable, a viewable page can be built up from sets of data derived from the teletext broadcast feed. One set, a skeleton page, provides all the static information. Another, the map page, is the definition of the screen layout and is used to control the organisation of data pages, the extraction of the relevant data and manipulation and display by computation within the workstation.

In the equities market a different situation exists. There is far more information available, most of which changes less rapidly, and the subset that a particular user wishes to view is likely to be unique to him and to change relatively infrequently. Here, it would be an advantage for the user to be able to design page formats at the workstation, provided this can be done quickly and in a clean and straightforward manner.

The user will be allowed to build map pages to describe screen layouts and store these map pages locally at the workstation. In order to build such map pages the user must be provided with a dictionary of the data that is available and with the ability to define the location and format of the data on the composite page.

At the simplest level it is only necessary for the user to identify the data items he wishes to view, define their location on the page and

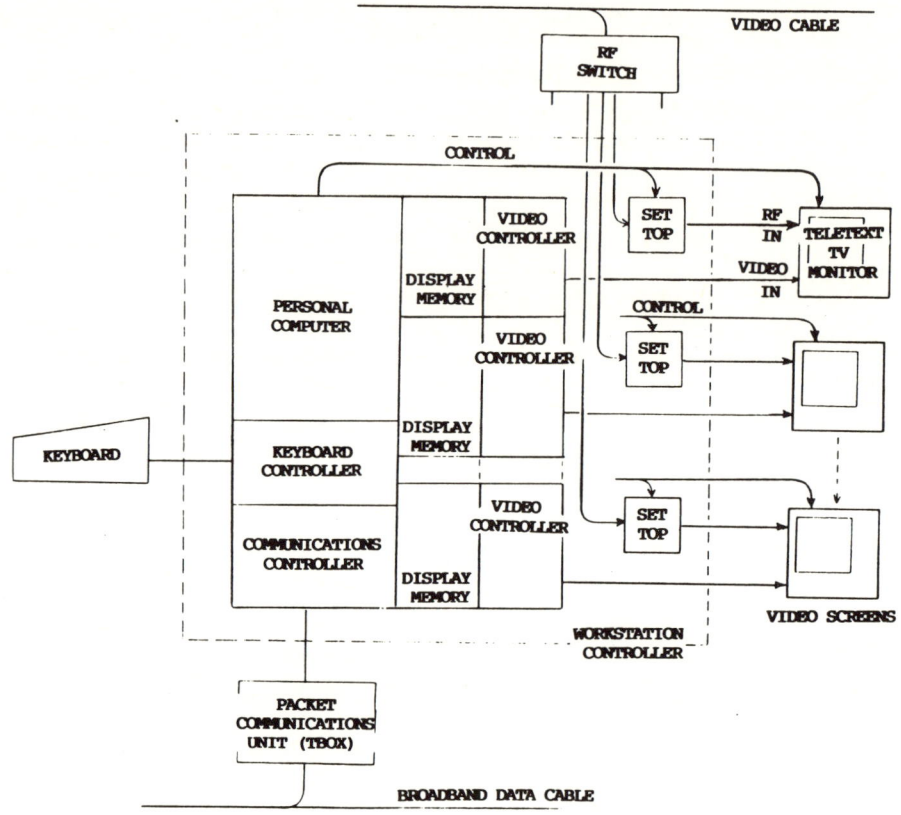

Fig. 17.8 User workstation.

add whatever headings and static text is required. Further enhancements would provide facilities for re-formatting the data, performing a variety of calculations on the data items and displaying the result.

Figure 17.8 shows a block diagram of a typical user workstation. The workstation will consist of the following major sections:

- Personal computer.
- Keyboard.
- A number of Full Field Teletext TV monitors.
- A set top convertor per TV monitor to provide the control and selection of TV channels via the RF switch.
- Display memory and video controller to generate the data display required from the personal computer.
- Communications controller to provide the interface with the packet switched network.

CHAPTER 18

Cable TV for an Integrated Services Local Network (ISLN)

In this chapter we will be looking at various aspects of providing non-entertainment telecommunications services on cable television systems. This is a relatively new area, and one for which no particular unique or standardised configuration has yet emerged. It has been shown that revenues due to such services can be significant and turn what might otherwise be a marginal financial proposition into a viable business venture[1]. However, the economics of cable systems which provide these services are of less immediate concern to us than the methods by which they may be implemented. The aspects covered are, therefore:

- The regulatory environment.
- Services.
- Implementation strategies.
- Some practical examples.
- Standards.

THE REGULATION ENVIRONMENT

The emerging regulatory scene in the UK

Over the last five years there have been numerous government reports, recommendations and legislative measures aimed at liberalising both telecommunications and broadcasting from the previous monopolistic or quasi-monopolistic structure. Such liberalisation, or deregulation as it is termed, is in line with a general trend towards meeting more closely the needs and expectations of consumers, not only in the broadcasting and communications fields, but also in many other facets of our lives. There were commercial pressures also – in the area of supply of subscribers' apparatus and maintenance services, and in the need to release growth and expansion in telecommunications from its dependence on government funding. This would permit consistent forward planning and encourage investment in a manner which would enable the demand for new and existing customer services to be satisfied.

The Telecommunications Acts of 1981 and 1984 therefore introduced measures to deregulate the provision of communications services by

- Allowing competition at all levels of service provision.
- **Privatising British Telecom (BT).**
- Creating a suitable regulatory framework with the establishment of Oftel, the Office of Telecommunications.

Of course, it is the competition aspect which is of most interest to cable system operators. Under the terms of the legislation, cable operators may offer a wide range of data and value added services freely, save in five city locations identified to date where such provision is seen as causing potentially unacceptable economic harm to the two main telecommunications carriers, BT and Mercury Communications. On the other hand, voice services can only be provided by cable operators in collaboration with either of these two organisations, at least for the moment.

Unlike deregulation in the United States, the legislation in the UK permits, and indeed encourages, competition in the provision of local network services. During its initial phase of operation, Mercury Communications currently intends to gain access to its customer base by the use of point-to-multipoint local microwave systems, operating at millimeter-wave frequencies in order to avoid interference problems. Such methods are economically viable only for a small customer base of relatively high capacity usage, and tend to squander otherwise valuable frequency spectrum resource. As the customer base widens to include less sophisticated users, Mercury will be forced to use more economic methods of accessing the subscriber.

One method which it is actively seeking, albeit with some difficulty, is by interconnection with BT's local network. While this has the advantage of almost universal connectivity, it suffers from a number of drawbacks associated with the quality and potential of existing and often old metallic pair cable plant. More significantly, perhaps, Mercury's dependence on BT for access to its subscribers cannot be said to represent real competition in the provision of local network services.

Cable television systems, on the other hand, represent an alternative and arguably superior means of providing such services. Not only is access to the subscriber completely independent of BT (except, of course, where BT is involved in the construction of a cable system), but the bandwidth and quality potentially available far outstrip that possible through the 2-wire loop, even taking into account narrow band ISDN developments. A cable system will also permit identification of a calling subscriber, an important feature of voice telephony interconnect not yet generally available in the UK telephone network. The use of cable television systems for providing non-entertainment services is therefore entirely consistent with the trends of deregulation which are taking place both here in the UK and in other countries.

Advanced interactive services

Notwithstanding the award of a franchise to operate a cable television system under the terms of the 1984 Cable and Broadcasting Act, it is still necessary for an operator to obtain a telecommunications licence, irrespective of whether he intends to provide interactive services. This is because the conveyance of television signals along a cable system is deemed to fall within the definition of a telecommunications service stated in the 1984 Telecommunications Act.

There are various classes of licence that may be issued. It is outside the scope of this chapter to dwell on these, but applications of sections of an Act can determine the nature of the licence to be awarded. One matter which has received much attention is the condition (or rather conditions) under which an extended 23-year licence may be issued as opposed to the standard 15-year licence. The conditions relate in part to the technical configuration of the cable system regarding the provision of switched television programming services. More significantly, they require that the cable system be capable either of carrying advanced communications services, or of being upgraded to carry such services within a given timeframe. 'Advanced communications services' in this context refers to the higher speed business services, possibly including digitised voice, for which special network configurations need to be devised.

SERVICES

Business and high speed data services

This categorisation, if a little awkward, corresponds both to a suitable market segmentation of services which can be made available over a cable system, and to a set of network solutions which facilitate their provision. There is obviously some overlap between the categories, at least in marketing terms, and this is illustrated in

Fig. 18.1, along with a listing of typical services for each category[2].

TELESHOPPING

Teleshopping is generally interpreted as the facility to conduct shopping transactions via the medium of a TV based service from the purchaser's own home. The intending purchaser is shown a list of the goods from which selection is to be made and can then indicate on his

	Domestic services	Small business services	High speed data and associated services
Teleshopping	X		
Telebetting	X		
Home banking/ cash management	X		
Home working	X		
Alarm and security	X	X	
Booking services	X	X	
Utility meter reading	X	X	
Energy and load management	X	X	
Information/ database access	X	X	X
Electronic mail	(X)	X	X
PSDN (and PSS) gateway access		X	X
PSTN access (voice telephony)		(X)	X
Video conferencing			X

Fig. 18.1 Categorisation of non-entertainment interactive services.

keypad those goods required, together with such relevant additional information as quantity, size and colour. The transaction is then completed with the sending of the order instruction. In the UK, this type of teleshopping does not require any direct human interaction; experience is limited to the various Prestel-based services available which range from purchasing relatively simple articles to full grocery ordering services. Two examples are Club 403 in the West Midlands ('Armchair Grocer') and Homelink in the London area ('Supermarche'). Links with mail order companies are also planned. With text-in-vision techniques, live demonstrations of the use of products are possible, and this aspect of advertising and promotion is particularly suitable for cable television systems where adequate video capacity is available.

TELEBETTING

Gambling in its various forms is still one of the most popular pastimes enjoyed by people in the UK and is also a significant consumer of the average person's disposable income. Betting on horse racing is already a popular service provided on the Prestel videotex service through Ladbrokes. Mecca have also been developing private videotex systems. Ladbrokes have, in fact, committed themselves to cable at an early stage through their involvement with the Ealing franchise.

HOME BANKING

The ability to undertake relatively routine banking transactions from home is gaining in popularity. The development of home banking in this country was pioneered by a group comprising British Telecom, the Nottingham Building Society and the Bank of Scotland. 'Homelink' was launched in 1983 via BT's Prestel Gateway service to enable customers to check their bank balance, pay credit card accounts, transfer funds, pay standing orders, and pay electricity, gas and other bills. The Bank of Scotland now operates full videotex-based Home Banking Services, and the Midland Bank has an experimental system.

The limiting factor in development of these services is the relatively small number of homes or even small businesses which have videotex facilities. There is, therefore, considerable potential for cable operators serving 100 000 homes to provide facilities for a bank to develop home banking within a specific catchment area.

HOME WORKING

There is an increasing tendency for people to work from home for a number of socio-economic reasons. The trend is perceived not only in self-employed groups, but also in employees of large companies.

Developments in telecommunication and distributed processing technology are clearly supporting a part of this activity, with transmission costs offsetting the savings in transportation and office overheads. The use of on-line computing facilities within a local call area is more widespread in the US, where flat rate charging applies. Cable television in the UK offers the possibility of gearing tariffs in a similar manner, or perhaps in relation to data volumes handled.

ALARM AND SECURITY SERVICES

A cable television system could facilitate the transmission of alarm signals from domestic and commercial properties, making possible swift and effective response from emergency services. Alarm indications may include warnings of fire, intruder, personal accident and medical alert.

The development of the alarm market has been hampered in the past by the inherent unreliability of signal transmission paths (typically BT land lines via auto-diallers), and by very high false alarm rates. A cable network can provide a more reliable and very probably cheap alternative link between users' alarm systems and the central alarm stations where the signals are processed. Cable operators could also operate a system of alarm verification, allowing the filtering of alarm cells before emergency services react directly.

BOOKING AND RESERVATIONS SERVICES

A wide range of booking services is possible including air flights, holidays, hire cars, taxis, concerts, theatres, newspaper advertising, etc. Many booking services are already available from a consumer's own home using a telephone to computer link (e.g. Teledata) or videotex system. Videotex systems are now used widely in the travel industry, a segment of the small business market which cable operators may wish to exploit.

UTILITY METER READING/ENERGY MANAGEMENT AND LOAD CONTROL

Telemetry systems may be used by the electric, gas and water supply industries for remote meter reading and implementation of tamper and leak detection facilities at customers' premises. These telemetry systems may be viable 'stand alone' systems or incorporated with a load management system. Load management is part of a demand management strategy and comprises any action taken by the customer or the supply industries to reduce the demand for supplies during peak periods. The deliberate reshaping of the customer's load curve can be achieved either by voluntary actions on the part of the customer or by actions imposed by the suppliers.

Load management communications can make use of cable systems – they are more reliable and interference-free than other media; the utilities' requirement for penetration to every customer's premises will, however, have a significant impact on the cable system configuration, and may require that a number of alternative transmission systems share the task of providing such high levels of penetration.

LOCAL AND NATIONAL INFORMATION SERVICES

The dissemination of local information to residents of a town is becoming a growth industry. All types of organisation – central government agencies (e.g. DOE, DHSS), various local council departments (e.g. libraries, housing, planning, PR), numerous voluntary bodies (e.g. CABX, community associations), and the media (radio and newspapers) – are becoming involved. National information database services to industry and business include financial, legal, political and economic data. However, domestic subscribers are also interested in information relating to sports, weather, travel and entertainment available nationally, and access to such databases via cable systems rather than, say, Prestel is commercially feasible. Of course, cable systems would be able to provide gateway access to Prestel (or to other, private videotex systems) as an alternative to the local PSTN, with competitive tariff structures which might attract wider usage.

ELECTRONIC MAIL

Electronic mail (EM) is a generic term describing any of a number of ways of sending text by electronic means. It therefore includes telex, teletex, facsimile, videotex and *computer-based message systems* (CBMS). Teletex is likely to have a particularly strong influence on the development of EM, especially in Europe. With the development of teletex standards and their adoption by a number of countries, the way is being paved for a global EM network. Value added services and computer-based message systems have played a significant role in the form of present day EM services and constitute possible opportunities for cable-based services. CBMS use 'mailboxes' for store and forward of messages. They evolved from computer time sharing systems, where users sent messages to other users on the same system. A number of systems have been set up in the UK over the last two years by companies already well established in providing EM services in the USA. Such services include BT Gold (ITT's Dialcom), Cable & Wireless's 'Easylink' and Istel's 'Comet' (from the Computer Corporation of America).

GATEWAY ACCESS TO PUBLIC DATA NETWORKS

The cable system headend provides an ideal point of concentration for data traffic originating and terminating within a franchise area. By establishing such a gateway, the cable system becomes, in effect, a wide area local network (WAN or sometimes MAN for *Metropolitan area network)* with access to the outside world. The WAN concept is, of course, also viable where there is a clear community of interest within the same area, for example a university campus, or several offices and plants of a large organisation situated within an industrial estate. Concentrating data traffic at the cable headend also allows multiple logical channels to be operated within the same connection to public packet switched networks. This can reduce access charges significantly, and bring packet switching within reach of the small user.

VOICE TELEPHONY

This term may be taken to include not only speech but also data services which are provided by voice channels. The present limit for these services is 19.2 kbps, but by use of a suitable PCM interface the data speed for one voice channel rises to 64 kbps. A cable system could give telephone service to a small percentage of its customers, without great cost or difficulty. These would be almost wholly business users, making or receiving long distance calls. Business users contribute over 80% of all telephone revenues, while representing only about 20% of the telephone subscriber population.

VIDEO CONFERENCING

This is a service which is expected to gain in popularity over the next decade in the face of escalating travel costs and an increasing awareness of the need for improved communications within large organisations. Its growth to date has been rather stunted, mainly because of high costs and inability to locate easily in customer premises. Cable systems, with their bidirectional video capability, can overcome these difficulties and offer video conferencing services at a cost which can compete with the alternative time and travel expense of meetings.

IMPLEMENTATION STRATEGIES

The types of services identified earlier may also be classified in terms of the data transmission speeds appropriate to their operation. This is illustrated in Table 18.1, with due allowance for the obvious overlaps that exist between the categories. Generally each broad category of service will lead to its own distinct network implementation solution, and we will examine the alternative configurations currently being explored by manufacturers and system designers. It is interesting to note that reference 3 suggests an alternative classification of services according to whether they are image, data or voice, and then relates these to corresponding modes of network implementation. This more generalised approach has the merit of including entertainment services in the set of network solutions.

Table 18.1

Range of data rates	Domestic services	Small business	High speed data services
1–100 bps	X	—	—
300–9600 bps	(X)	X	(X)
9600 bps to 64 kbps	—	—	X

Domestic interactive services

In dealing with widely disseminated, very low speed domestic services, it is not economic to employ the serial data protocols and modems used at higher data transmission speeds. Moreover, the fact that many of the services in question make use of the domestic television for display purposes indicates that use of a modified form of teletext or videotex would be appropriate. A number of switch manufacturers, Thorn EMI and Rediffusion among them, are using this approach in their systems, for both internal network control and interactive subscriber services.

It will be recalled that teletext makes use of the vertical blanking interval (VBI) between successive fields of the television signal. Because the lines on the vertical extremities of a television picture are not usually visible, they may be used to convey data bits in place of the usual chrominance and luminance information. The information contained in these lines comprises the page address, display location data and the displayed character data itself. (There are also redundant bits for synchronisation and error control.) Of particular interest is the page addressing capability which can be adapted in cable systems to identify uniquely each subscriber. Seven digits are available, the first three of which are used in traditional broadcast teletext. (This is because the number of pages available on a broadcast system is constrained by a required mean access time of about 15.) By judicious use of both TV set and cable system keyboards, elementary domestic interactive services may be directed exclusively to the subscriber who requests them.

A typical sequence is for a subscriber to enter prompts or a security number which are held on the keypad until polled from the star switch. The downstream responses may comprise bursts of display data occupying one or two fields of VBI teletext. Since these are transmitted once only to the TV receiver's page memory, the opportunity for eavesdropping on the downstream channel is remote. If required, additional security protection can be arranged by formatting pages of sensitive information (e.g. bank statements in the Home Banking Service) in an anonymous way.

Interactive services may be extended to subscribers who do not have teletext-adapted TV receivers by providing decoders and pixel stores in the star switch. These would be provided on a contention basis of say one per ten subscribers. This approach is also used in the GEC switch design, although full videotex rather than modified teletext is proposed for implementing domestic non-entertainment services.

Medium speed business data services

Small business data applications will require data transmission speeds of 300–9600 bps. The bandwidth occupied by a population of simultaneous users is significant, but can be contained within the capacity of a cable television system by a suitable combination of techniques.

TOPOLOGY
While it is possible to harness bi-directional data communications on to tree-and-branch type cable networks, the problems of upstream noise accumulation require complex solutions. One technique, developed in the US and known as 'Homenet', partitions the cable franchise area into cells which are apportioned a segment of data communications capacity[4]. Conceptually this may be seen as converging towards a hub or star configuration for data communications purposes. The problem of adapting existing networks did not arise in the UK, and the opportunity to introduce switch star technology with its attendant flexibility was taken.

Equally important is the issue of network control, which in a tree-and-branch structure is inevitably more centralised and complex in an area covering 100 000 homes. Star switch topology enables intelligence to be distributed throughout the network at the star points, where a number of steps are taken to simplify the problem of network management. These include traffic concentration, local subscriber processing and dialogue with the headend control computers.

BROADBAND LAN

The star switch topology allows controlled access of subscribers to logical channels of a broadband local area network. 'Broadband' is simply a term describing cable systems in which the transmitted information (e.g. data generated from users' terminals) modulates analogue RF carriers, each of which occupies a given channel of the frequency spectrum. This frequency division approach differs from the so called 'baseband' LAN in which such data is transmitted directly to the cable without modulation. Broadband systems are more flexible in the way different services may be handled, since there are usually many logical RF channels available, the bandwidth (and frequency) of which may be adjusted in accordance with the application or service to be carried. They are also capable of extending over the area or length of the entire cable franchise area to become, in effect, a wide area (WAN) or metropolitan area network (MAN).

PACKET SWITCHING

A packet approach to transmission on cable systems is consistent with the 'bursty' nature of data originating from terminals, and with CSMA/CD protocols for sharing logical channels of a broadband LAN. Such protocols have the added advantage of providing inherent error detecting and error correcting properties, particularly if an HDLC-type protocol is used. In contention type protocols, it is difficult to differentiate a transmission error from a collision. Recovery can be achieved in most packet protocols by a repeat of the packet transmission. The packetising of data in the cable system is also convenient for connection to external public data networks (PSDNs) which are also generally X.25-based packet switched networks. The gateway for such interconnection, located at the cable head, can be arranged to minimise the number of total connections to the PSDN, making concentrated use of the available logical channels per access port. The resultant savings in PSDN connection charges may be partially offset by the need to provide some addressing overhead and control at the headend.

SPECTRUM ALLOCATIONS

Institutional broadband networks often consist of two cables, one for each direction of transmission. Alternatively, a single cable may be used in a virtual 4-wire mode, where each direction of transmission occupies half the available spectrum. In cable television systems, data transmission capacity is allocated a much reduced part of the band, the lion's share of spectrum being devoted to television entertainment services. In switched star systems the trunk cable spectrum must be well planned if all potential demand is to be accommodated without the need and expense of providing additional cables. The subscriber drop cable, on the other hand, has more than adequate capacity. Most trunk cable systems in the UK use a mid-split approach in which the band between 0 MHz and 50 MHz is assigned to the upstream direction, and that between about 80 MHz and 450 MHz to the downstream direction. To this extent the word 'mid-split' is something of a misnomer, probably derived from the term 'sub-split' used to describe the frequency division at 30 MHz typical of North American tree-and-branch networks.

Related to the question of frequency allocation are the issues of spectrum utilisation efficiency and modulation methods. There is considerable latitude on the subscriber drop cable and the simplest of RF modulation schemes may be employed. These range from simple on-off carrier keying to 2FSK methods, depending on data rate. Neither of these is terribly robust in the presence of noise, but we assume that good system design can ensure that this need not be a constraint – the cable system constitutes a bandwidth-limited situation rather than a power-limited environment. Two-level FSK coding is inefficient in terms of bits per hertz (0.1–0.3 bits/Hz) and for higher speeds (0.5–2 Mbps) a more elaborate scheme is necessary, for example quadrature phase shift keying (QPSK) or even 16 amplitude phase shift keying (APK) schemes which give efficiencies in the order of 1–2 bits/Hz. Such higher speeds are relevant to the business data market because of the possibility of data concentration at the star point. Multilevel encoding is only possible at the higher speeds because of the expense and complexity of the relevant RF modems. FSK has been a widely used and thus cheaper modem technology for some time.

Two further complications of RF modem design for cable systems are worthy of note. First, there is a need to ensure good spectral purity to avoid out-of-band emission, and efficient screening to safeguard against ingress interference. Such screening is required in all parts of the cable plant in order to avoid the problems of impulse noise and associated burst errors. Secondly, included in the whole area of network control and maintenance is the notion that faults in the cable system components should be able to be isolated and differentiated from those in subscribers' apparatus. Facility for automatic local loop-back testing should, therefore, be included in the modem design.

High speed data and voice telephony services

Large business users – and for that matter domestic users – with a requirement for voice telephony and high speed data communications impose a much more stringent demand on the cable system in at least three areas:

• Trunk dimensioning.
• Network topology.
• Unit cost of service provision.

The latter point will only be solved with high volume manufacture and standardised products – an issue explored later. With regard to trunk dimensioning, it is not possible to assume that multiple simultaneous sessions of voice or high speed data traffic can co-exist with multi-channel TV transmission on the same cable. This makes it necessary to consider additional 'overlay' cables, and to devise the interrelation of such additional cables with the existing entertainment cable topology and configuration.

A further related problem is that star switch or more accurately 'mini-star' nodal structures constitute the focus of too few subscribers to be of any real value as a contention point for the concentration of telephony traffic. In telephone networks, local exchange capacities of 1000–20 000 lines illustrate the economies of scale needed for efficient traffic concentration and switching. The network designer must, therefore, attempt to configure the telephony overlay in a manner which maximises the opportunity for traffic concentration.

Much will depend on the density and extent of coverage proposed for telephony and high speed data services. If service to a small pocket of subscribers only is envisaged, say in an industrial estate, a complete overlay to subscribers' premises may be the most economic solution, especially if digital PABX trunk interconnections are required by many users in the area. These involve 30-channel groups operating at 2 Mbps, for which RF modem technology is relatively mature and cost-effective. Lower capacity or even single 64 kbps voice channels might use RF modems of similar technology and design, resulting in a very high cost per circuit. In that event, it may be better to adopt a simpler strategy for connecting single streams of high speed data or voice. One way is to use the coaxial subscriber drop cable as a straightforward 2-wire loop, akin to the symmetrical pair cable used by BT. PCM encoding could then be carried out at the star point or beyond, using standard primary multiplex equipment. As a further measure of conserving bandwidth, demand assignment of working channels on to the trunk cable could be investigated, using techniques similar to those already applied in satellite and cellular radio systems. It is considered unlikely, however, that such techniques will obviate the need for trunk overlay cables without unacceptable sacrifices of capacity on the entertainment cable. At a later date, when ISDN standards and equipment are widely available, the subscriber drop cable would provide a high quality (and somewhat simpler) 144 kbps (2D + B) narrowband ISDN channel, or even a 385 kbps wideband ISDN path when standards for this are developed. Unlike the former, wideband ISDN channels would not be capable of being carried on the traditional symmetrical pair subscriber loop.

SOME PRACTICAL EXAMPLES

Cabletime approach[5]

Simple signalling or voting is possible using the keypad and set top unit in the Cabletime 16 mini-hub system. However, the provision of interactive services is centred on the medium speed business data service category which, it is claimed, can also be configured to serve domestic users economically. The design is based on use of the Sytek LocalNet 20 broadband LAN provided in collaboration with ITL, Sytek's UK distributor, under the trademark Cablestream.

Spectrum is assigned on both drop and trunk cable (see Fig. 18.2) to LocalNet 20 FDM channels, each of which operates on a CSMA/CD packet switching basis at speeds up to 19.2 kbps. Access to data channels is made via *network interface units* (NIUs) which are in effect local PADs (packet assembler/dissassemblers) which format serial data from subscribers' terminals (RS232C-CCITT V.24) in

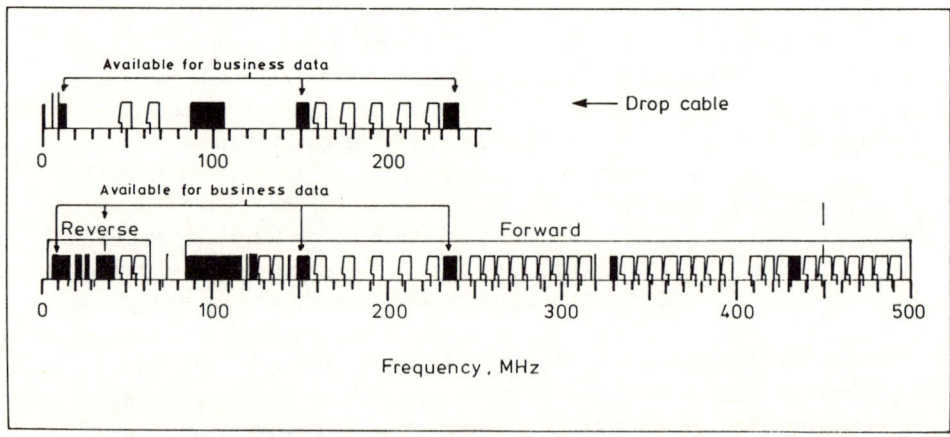

Fig. 18.2 Cabletime system spectrum.

accordance with the CSMA/CD line protocol. A frequency translator at the headend connects upstream to downstream channels to make up transmitter/receiver pairs. Sixty-channel capacity at an aggregate data rate capability of 128 Mbps per channel is provided on the trunk cable spectrum. Domestic data services at say 300 bps may be provided economically, it is claimed, by locating the NIU at the star point on a traffic contention basis, permitting it to be seized by a requesting subscriber, who may use a simple FSK add-on modem for transmission over the subscriber drop cable. A schematic of the arrangement is shown in Fig. 18.3. This diagram also shows a possible arrangement for voice telephony and high speed data by the provision of 30-channel PCM primary multiplex equipment at the star point, feeding an overlay cable to the headend gateway switch.

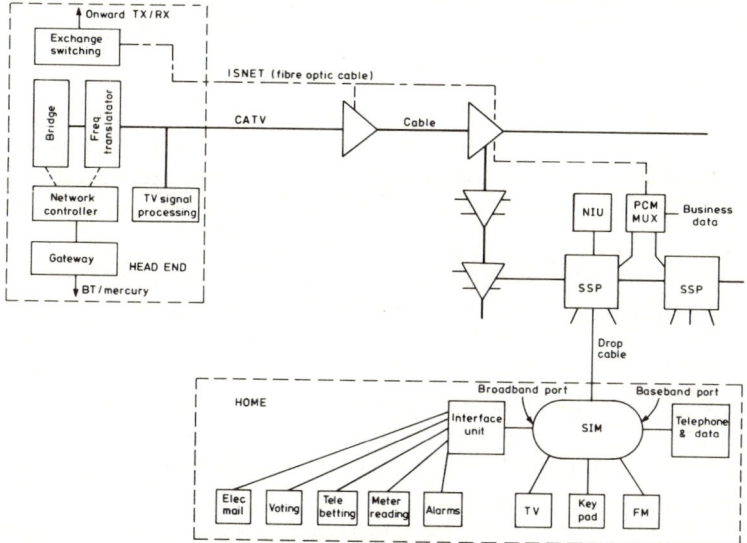

Fig. 18.3 Cabletime network configuration for providing interaction services.

The overlay cable may be coaxial or optical fibre. Cabletime's ISNET (Integrated Services Network) concept of optical fibre pair trunk overlay and baseband drop cable connection to the subscriber is based on a future scenario of providing high speed data and voice service with 144 kbps ISDN channels to every home.

Thorn EMI approach[6]

As stated earlier, the Thorn EMI system makes use of quasi-interactive teletext for the provision of domestic non-entertainment services. Keypad entries are carried to the switch at the rate of 1 byte every 160 ms or 6 bytes/s. In the downstream direction, full field teletext is decoded at the switch and converted to VBI form for transmission to subscribers' teletext TV sets, or to monochrome video frames for subscribers without teletext adapters. Modulation on the drop cable is 2 FSK.

For business data, a trunk cable spectrum allocation of 18 MHz in each direction of transmission is made available (see Fig. 18.4), based on a mid-split (or as Thorn EMI term it more accurately, a quarter-split) cross-over point somewhere between 56 and 80 MHz, depending on the directional filters employed in the line system. Like Cabletime, Thorn EMI intend to use the Sytek broadband LAN approach for general business data services. They also plan to have an optional 64 **kbps full duplex modem adaptor working in the band 5.5–6.0 MHz for** software downloading, business use and computer communications.

STANDARDS

The significance of standards for wideband cable systems

There are a number of convincing reasons why the development of standards for wideband cable systems is important to the industry. First, the fragmentation of regions of the country into individual and independent franchise areas means that there will be many diverse cable system operators. Each of these will be carrying out sovereign decisions as to procurement of sub-systems and equipment within their territory, without necessarily consulting others in the process. There is a danger, therefore, of a proliferation of different network types and configuration which may result in high unit costs, not only in terms of manufacturers' prices, but also in respect of the engineering effort such custom-made solutions will require. Standardisation of network solutions would avoid this problem, and simplify the job of system design and construction for all operators.

There is a second, more strategically important issue at stake. The viability of cable as an alternative to other means of providing local network services rests on finding cost-effective methods of implementing non-entertainment services. The wideband capabilities of coaxial cable are being challenged by such developments as ISDN standards for symmetrical pair cables. The technological inertia of the current PTT telecommunications networks means that these developments will take many years to mature into reality. There is, therefore, a clear opportunity for wideband cable systems to capture this segment of the market, provided that the unit costs of associated

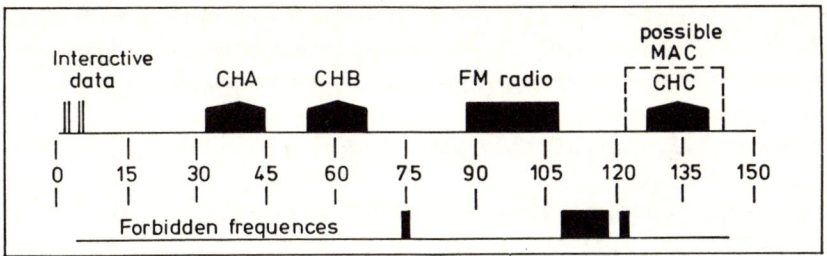

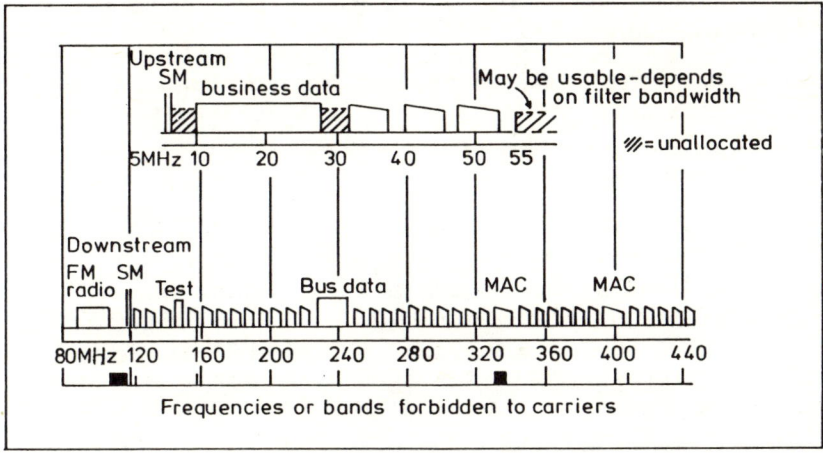

(a)

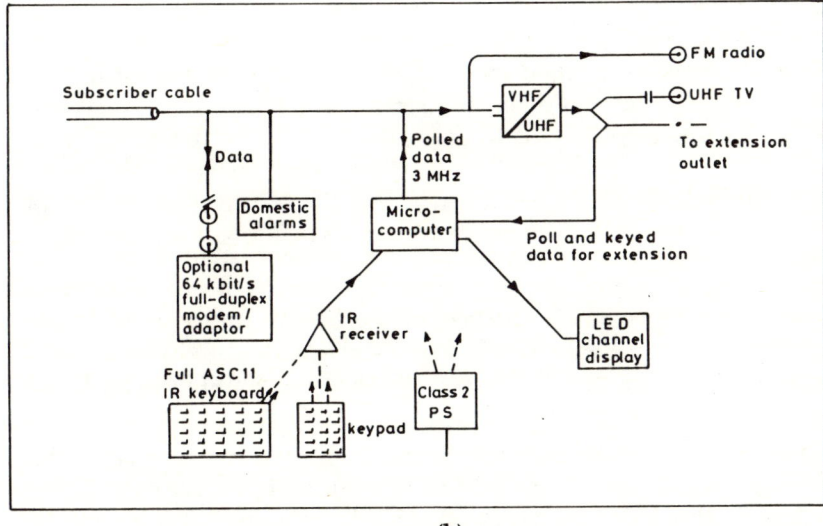

(b)

Fig. 18.4 (a) Thorn-EMI system spectrum; (b) subscriber interface in the
Thorn-EMI system.

equipment can be brought down to acceptable limits within a relatively short timeframe. The existence of firm standards for network configurations and interfaces will facilitate this process.

The Eden Committee standards work[7]

The Eden Committee was established in 1982 by the Department of Industry in order to set in motion the BSI procedures by carrying out preliminary work on standards for non-entertainment services on wideband cable systems. For each of the services considered, the Committee considered:

• Interface between the cable system and external networks to which it would interconnect.
• Interface between the cable system and subscriber.
• Subscriber addressing.

In general, all standards work involved relevant CCITT X series and related recommendations. In the case of the interface to external networks, the Eden Committee proposal was to propose the CCITT X.25 Recommendation as the standard between the cable headend and the point of interconnection with BT's (and Mercury's) public packet switch data network at data rates of 2400, 4800, 9600 and 48 kbps.

Interfaces between subscribers' terminal equipment and the cable system were generally based on CCITT Recommendation V.24 (RS-232C), although special arrangements, especially at the higher speeds, between cable operator and subscribers were not precluded. Both asynchronous and synchronous working were addressed.

The main services studied by the Committee were those of teletex and videotex, the former seen as primarily a business service, the latter with some application to domestic interactive services. With respect to external networks, teletex was seen as a value added service carried on the PSDN on an X.25 basis. A number of alternative scenarios for the provision of service to subscribers were studied, including:

• Direct access through PSDN line extension by the cable system.
• Headend PABX links to teletex terminals.
• Headend cable mail server supporting personal computer type user terminals and standard teletext TV receivers.

With regard to videotex, several methods of delivery service to the subscriber were studied, in both broadcast and interactive modes and according to the different types of subscriber terminal device to be used.

REFERENCES

1. Snoddy, R. 'Why Entertainment Alone Will Not Produce the Profits' (1984) *Financial Times*, 9 April.
2. 'Provision of Non-entertainment Services on Broadband Cable', *Multi-client report prepared by Cable Systems Development Company Limited*, in collaboration with Ewbank Preece Management Limited.
3. Willson, R. *A Strategy for the Provision of Interactive Services via Cable TV Systems* (1985) Presented at *Cable '85*, Brighton, July.
4. Maxemchak, N. F. and Netravali, A. N. 'Voice and Data on a CATV Network' (1985) *IEEE J. on Selected Areas in Comm.*, **SAC-3**, 2, March.
5. Watson, L. 'Planning for Efficient Broadband/Cable TV Networks' (1984) *Comm. Eng. Int.*, October.
6. Barnes, P. 'Cable TV in the UK' (1985) *Electronics and Power*, June.
7. Dawson, G. 'Standards for Interactive Services on Wideband Cable Systems' (1985) Presented at the *Int. Conf. on the ISDN and its Impact on Information Technology*, IEE London, January.

Index